Homo-Erectus as a Man

By Steve Preston

1st Edition

© Copyright 2017, Steve Preston
All rights reserved.

Contents

HOMO-ERECTUS AS A MAN 1
INTRODUCTION 5
OUT OF AFRICA 9
SECRETS OF DNA 12
DATING PROBLEMS 22
GIGANTUS TO MODERN MAN 28
HOMO-GIGANTUS BEGINNINGS 36
ANAK CIVILIZATION 50
DINOSAUR FUN 70
CURIOUS CHARTS 74
END OF THE ANAK 78
HABILIS-LIKE APE-MEN 80
HOMO-ERECTUS NALEDI 82
HOMO-HABILIS APE-MAN 84
HOMO GEORGICUS 88
ERECTUS COMPARED TO HABILIS 91
ERGASTER ERECTUS 93
ERECTUS MUTATION 96
RHODESIAN MAN 99
SE ASIA MUTATIONS 101
ANTECESSOR ERECTUS 104
IDALTU ERECTUS 106
HEIDELBERG ERECTUS 108
NEANDERTHAL ERECTUS 111

ANAK MOLDED NEANDERTHAL	115
TESTING NEANDERTHAL DNA	120
DENISOVAN ERECTUS	126
FLORESIENESIS ERECTUS	132
GRIMALDI ERECTUS	134
BOSKOP ERECTUS	137
CRO-MAGNON RACE [FM:NF]	139
WAR AND EXTINCTION	146
CRO-MAGNON OF GENESIS	151
Y-DNA HAPLOTYPE TRACKING	162
MITOCHONDRIAL HAPLOTYPE TRACKING	166
HAPLOTYPE AND RACE	175
SHORT BIBLICAL RANT	177
MORE MUTATION	191
VANARA PEOPLE	194
CHIMPANZEES	197
REFLECTION AND CONCLUSION	204
ABOUT THE AUTHOR	208

Introduction

Have you noticed that people have not mutated significantly since before historical records was introduced? There are no new "races" or "hair color" or anything so one might wonder how did we get evolved? Now think about the Law of Entropy which essentially states that in nature everything devolves. Evolution would always be to the less survivable which would not survive. This is not a theory, it is considered a law so how did we "evolve" from Homo-Erectus and how did Erectus evolve from Homo-Habilis and all the predecessors? In the classroom you were taught about the Australopithecines and Homo-Habilis variants as if they were people. They were not; but while Homo-Habilis was roaming the African continent; Bam! Homo-Erectus miraculously appeared with a brain 50% larger and his characteristics changed dramatically. Erectus was designed differently. A new hip that could walk as easily as modern man popped into existence and the placement of the skull changed to allow him to walk and run. To make this even more exciting we are told Erectus could sort of talk. All this advancement does not mean he was the only human on the planet, in fact, massive blackened bones have been found in North America of an even older species of human. The discovery would not be localized, but somehow, this is also not typically taught.

In this book we will investigate how "erectus" got here, what he was like, how he fits into the real timeline of human development, the strangeness of Bonobo and Chimpanzee, and how the Erectus fits in with the discovery of giant men walking with dinosaurs and that whole mess. We will travel with them as they are modified and we will review many texts that describe how that came about.

You were told about the variants of Neanderthal, but you were not told about the relationship of Erectus to this anomaly of a man and how Homo-Neanderthalis discredits that whole "Out of Africa" theory that tied up human evolution into a neat little bow. If that omission is OK with you, I would suggest you not read this book. If you yearn for truth, I think you will be surprised by what scientists tell us about DNA, evolution, and how we got here. Don't get me wrong, there are many studies out there and it seems like many contradict each other, but more and more they are converging on a more probable history. What you probably learned in school was something called the "Out of Africa" development of mankind so we had better look at this theory that Homo-Erectus popped out a couple million years ago [100-thousand years ago by more accurate timing] and populated the world. Before you think it's going to be an easy road look at the graphic following. It is not that straight-line evolution you're reading about. Most human types simply vanished.

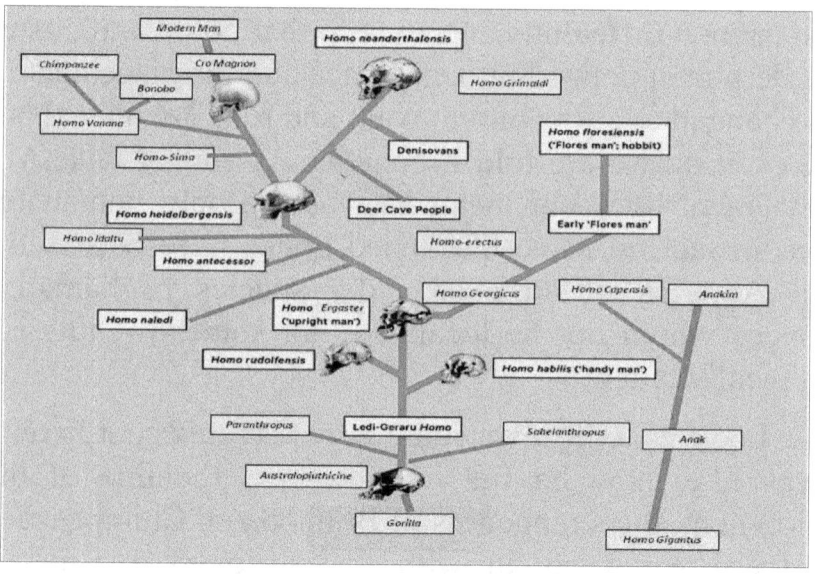

Here is the funny part. DNA scientists know all the areas of concentrations for various DNA mutation and know approximately when each mutation occurred, but trying to force fit desired outcomes has made their human tracking product a mess. If we just look at a few; the first one following shows the HLA gene that protects us from diseases came for the Orient and hasn't completely gotten down to the bottom of Africa yet. They try to tell us this mutation was very recent and after dissemination of populations around the world without explaining why Africa was left out. The second one describes Neanderthal separating from African populations 600 thousand years ago but all modern humans coming from Africa separately. The third shows Denisovan were the ones leaving Africa early with Neanderthal staying in Africa until 60 thousand years ago with no interconnection Denisovan. On the bottom row we first see modern humans lived in Africa until 100 thousand years ago, left no skeletons and all of a sudden Cro-Magnon skeletons showed up in the Middle East and Europe. The middle chart shows Denisovan and Neanderthal separating 80 thousand years ago and with no apparent contact and both mixed with Orientals by some unknown process. It also has Floresienesis not coming from Neanderthal variants when others try to show that link. The last chart indicates Denisovan linked up with an unknown group with different DNA and while Denisovan and Neanderthal came from the same DNA group, they bonded again in the near term.

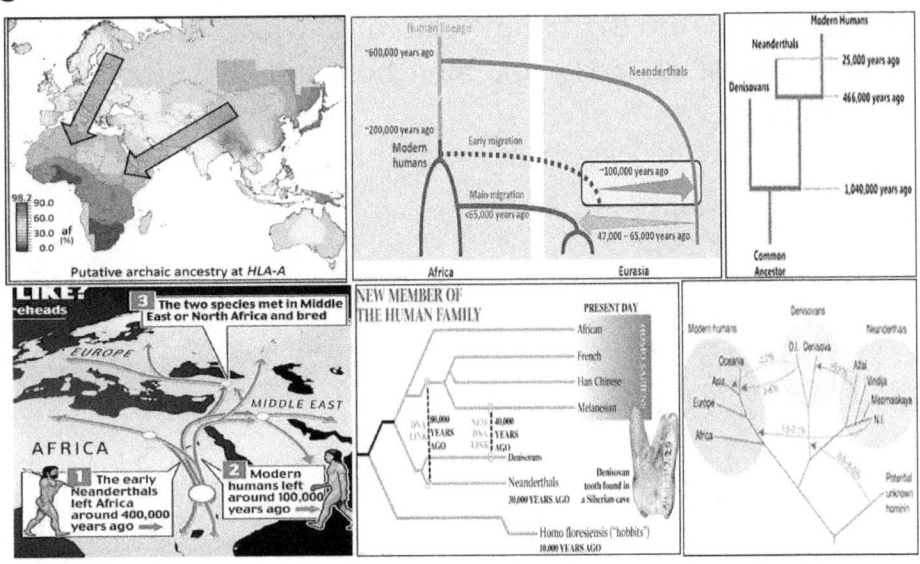

Yes these are all messed up in one way or another, so how can you get a clear picture about what happened if the Anthropologic scientist can't agree on ANYTHING.

- Quit ignoring the Anak people and the fact that they had an advance culture with high levels of technology.
- Quit ignoring the capability to modify DNA during the Pleistocene Age that has been proven over and over.
- Quit trying to eliminate the Pleistocene Extinction and worldwide flood that restarted civilization and spread clusters of civilization to other area as weather changes were significant.
- Quit ignoring the "Bharata season of War" from 3500 to 3100 BC and its nuclear probabilities during the time that massive human mutation occurred.
- Don't be afraid to review ancient details locked in historical references including the Bible. For fear the outcome may have a hint of religious acceptance.
- Quit trying to rely on evolution as the only solution to solve development mysteries when it is one theory that has done exactly the opposite causing many anomalous questions to ruin our history.

Quit trying to force fit the "Out of Africa" mantra simply because almost all early ape-man hominids have been found there.

Out of Africa

For whatever reason, scientists have been trying to force fit something called "Out of Africa" to show modern man originated in Africa. They even made up complex maps showing how all of this happened, as shown below. The dark blob in Africa was supposed to be how Homo-Sapiens "evolved" in Africa and stayed there of years before finally venturing off to the Middle East and then to the rest of the world. If you notice the arrow from Asia to North America, that was a lie as well.

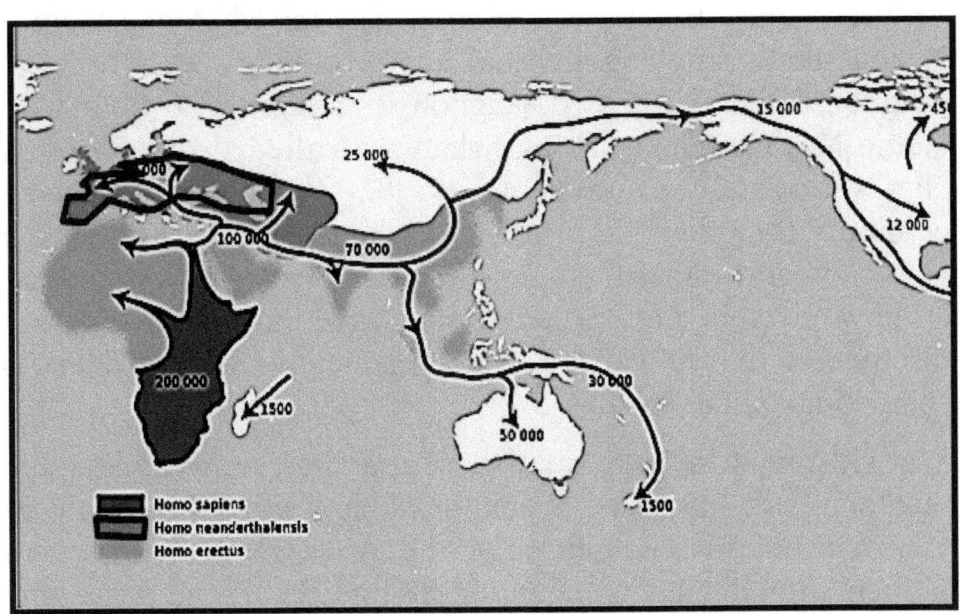

The problem is; none of it happened. While the actual spreading of Homo-Erectus and Homo-Neanderthalis has been debated, we know where we found fossils shown below. Left is primitive Erectus and right is Neanderthal Erectus. Please notice they developed separately. This includes sexual encounters.

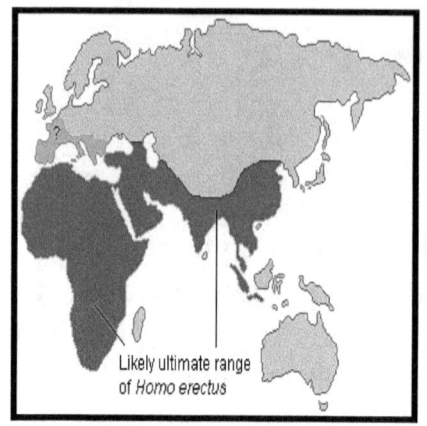
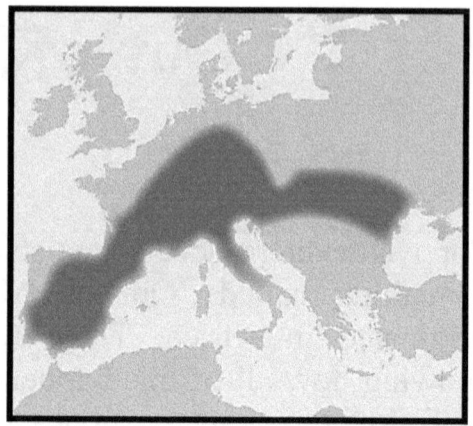

Then when it came to Modern Man starting with Cro-Magnon, the first known fossils were from the Middle East and Africa people did not mix with Cro-Magnon offshoots for many years. Rough numbers are that there is a 4% similarity between Homo-Erectus and Homo-Ergaster DNA and Homo-Neanderthalis showing a casual inter-grouping well after Cro-Magnon came along. Cro-Magnon is dissimilar to both with Homo-Neanderthalis having what are called "Alien" genes. His was distasteful to some as the African people were certainly the oldest, but their heritage was derived from Homo-Erectus rather than Cro-Magnon. We are going to look at how all this got turned around and how scientists are discovering the truth. The main component in this discovery is something called DNA.

We will review a large number of graphics, maps, charts and diagrams and introduce a couple of new groups to some of you. The Homo Gigantus and the Anak people were very influential in the development of Homo Erectus and those who followed so we must also learn about them a little more. While many bio-geneticists try to ignore these first groups, we will need them to understand more about Paranthropus, Sahelanthropus, Rudalfensis, and Georgicus ape-men who were all considered Homo-Habilis or his immediate predecessor. After the Habilis we come to the group we classify as Erectus including, Ergaster, Erectus, Rhodesiensis, Antecessor, Idaltu, and Heidelberg. Of this group, the Heidelberg gives people the most problems so we will certainly look at how that group plays into man's development. That is followed by the Neanderthal cluster including Denisovan, Floresiensis, Grimaldi, and Boskop Man. For this book we will look at this secondary group or offshoots of Homo-Erectus and explain why they all should be considered Homo-Erectus. All of a sudden everything changed as a new human suddenly appeared and was called Cro-Magnon by scientists. He was larger, had a larger brain, and generally he was more advanced than modern man as the following 20 thousand years of mutations began reducing our capabilities; following the Law of Entropy as all uncontrolled evolution does. Finally we find the Vanara people and Chimpanzee who evolve from man. We will have to look at the details of this new revelation simply because it seems so very strange. I know it sounds like a long list to go through, but I guarantee you will learn something and have fun at the same time gaining details many struggle with.

How people survived the Pleistocene Extinction, Earth Axis shift, and worldwide tidal waves and flooding- We will get into this in more detail later, but many survived the extinction, 10 thousand years ago, and re-located as the Earth's temperatures went crazy and places that had been warm were now freezing. As the water level was 400 feet higher after the South Pole settled on top of Antarctica one could not easily get to Australia and rivers had turned to Seas. By ignoring this critical element of man's development, wrong assumptions produced wrong answers. One thing that is helping us correct a number of issues is something called DNA mutation.

Secrets of DNA

Have you ever wondered how these "I-can-find-your-ancestors" companies work their magic? Part of it is done with an alien. I don't know if you know it or not but there is a tiny alien living inside each of your cells called mitochondria. At some past time, it may have been harmful to us, but now our cells use mitochondria. For scientists they are interested in the Mitochondria's DNA as it is different than your real DNA. Mitochondrial DNA mutates differently than does our "own" DNA. Here is the really neat part---SPERM. Sperm, like all the other cells in your body have DNA in its nucleus and another set of DNA in the mitochondria. As it swims to be the first into an ovum, if successful, the head easily penetrates the egg wall, but then something bad happens, the egg's outer shell becomes hardened and no other sperm can enter. This happens so quickly that the Sperm tail is amputated and that is where its mitochondrial DNA is, I mean was, located. Let's have a moment of silence for the sperm tail. The images show the first sperm entering and then the swarm unable to penetrate. Somewhere in the mix is a tail floating by.

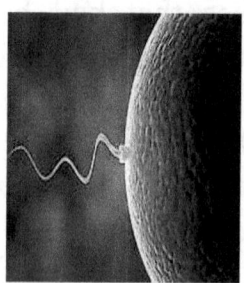 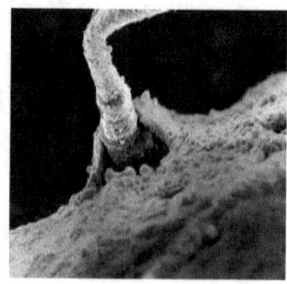 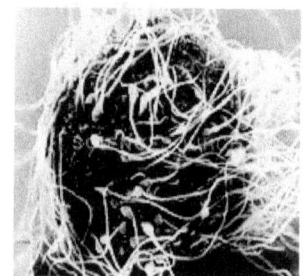

Why is This Important?

Because we keep losing sperm mitochondrial DNA but carry the female mitochondrial DNA we can see mutations of both DNAs and if we are testing the mitochondrial DNA, we know it comes from ONLY female ancestors while the "normal" body DNA called "nuclear DNA only traces family generations [male and/or female]. Many times the nuclear DNA is obtained from the "Y-Sex" Chromosome in a cell. A description of this detail is shown below.

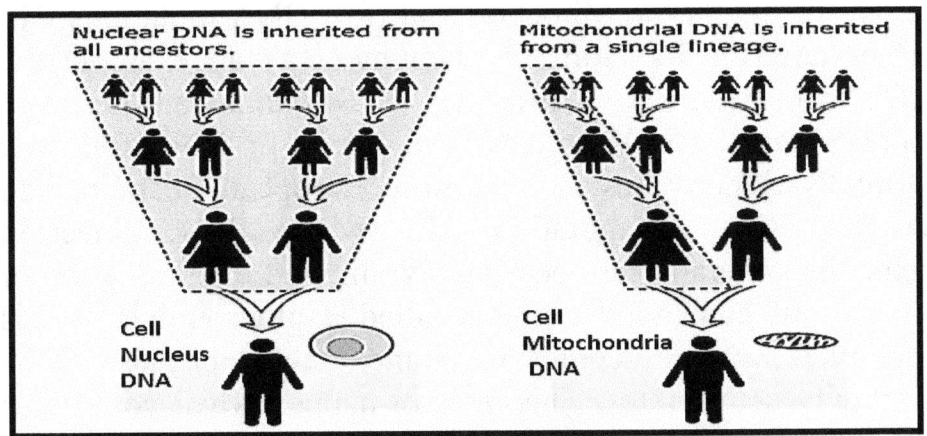

Mutation Memory

DNA is interesting in that it has a memory of mutation. Each time it mutates, the "scar" of that mutation is placed in the chromosomal string as a specific change in one of the sugars making it up. As more and more scars are built up a record of when each mutation occurred and how serious the mutation was. We will mostly look at the really serious body changing mutations to find out more about Homo-Erectus. As you can imagine there are many more major mutations in the "Nuclear" side as the record passes through more individuals, but we find that family modifications can be quantified into smaller genetic mutation so we actually have a smaller set to review. This will make sense as we go along. The most serious mutation is depicted as a capitol letter followed by a less serious one as a number and then comes a lower class letter followed by a letter. Once the coding gets too complicated, additional codes were made for "groups". Let me give you an example. [R1b] this is the Y-Chromosome mutation associated with Europeans. Depending on where on the Y-chromosome this set of mutations is located tells a researcher when a person came in contact with a European, how many times and the timing of each of the various mutational groups interacted with the test subject. By simply counting the number of [R1b] groups one can tell what percentage a person is with respect to Europeanism and the Mitochondrial [mt] DNA mutations will further describe locations of encounter. As an example the mtDNA mutation "V" indicates a Scandinavian heritage. Therefore an [R1b$_m$:V$_f$] mutation signature [Y-DNA followed by mtDNA] describes a more pure European Scandinavian than an [R1a$_m$:V$_f$] mutation. This whole Bio-Engineering science is generally termed Haplotyping and determination of how a group comes into being is called Haplogrouping. This is incomplete when it comes to Homo-Erectus because DNA eventually deteriorates. This doesn't happen when you die as your DNA after death is the same as DNA when you were alive but after 50 thousand years or so many of the DNA sugars are broken down and the history becomes incomplete.

As DNA doesn't lie [often] we can get a better understanding of how people got here than listening to someone spout off about all humans coming out of Africa because "Lucy" the Australopithecus Ape-woman was found; the first Homo-Habilis; and then the first Homo-Erectus was found there. Some might stupidly think the rest of the world was devoid of anything more advanced that a worm, but that is simply not the way these things go.

If you noticed I did put the caveat [often lying] as people can still manipulate things. Such is the case of the first Haplotype family tree of major mutation in the Y-Chromosome. It is shown next. The problem is that the "CF" dual mutation never happened as F was completely separate and found in the

Middle East while C, along with A, B, and D all were placed in Africa. We also now believe C mutation humans did not venture out of Africa, but they seem to have come from Europe; and there is a line to Homo-Neanderthalis very late in development time, so the whole thing is skewed just to make it look like people all came out of Africa. By the way; this chart and many like it concentrate on the major mutations. [Capitol Letter]

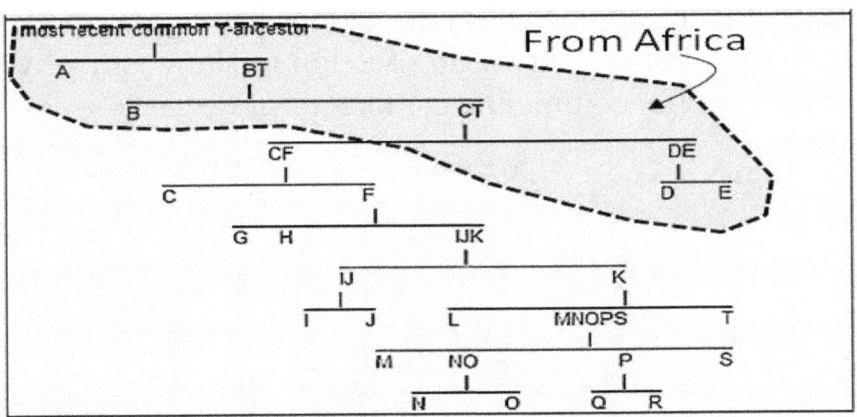

We will correct this thing as we go along so you can appreciate man's development and how Erectus really played into modern man.

Another Race of People

To understand Erectus and Homo-Neanderthalis we need to understand the "other" human race. Greeks called them Homo Gigantus or Titans, but the Bible simply called the Giants of old [Genesis 6:2]. How these first people became the group known as the Anak [Jewish], Annunaki [Sumerian], Akamim [Mongulala of Brazil], Lord of Amenti [Egyptian], Archaics [Adena of North America], Araya [Dravidian of India] will be reviewed briefly so that you can appreciate a truer history, truer religion, and truer science than what has been spouted off by teachers all over the United States. We'll travel through time from the Cretaceous up to the end of the Pleistocene with stops along the way to view what is happening to mankind before Cro-Magnon came along and changed the life of men forever. A brief timeline of some of the more well-known modified Apes, Homo-Erectus, and the modification of that special man are shown below. Some of the dating is changed in the chart since the radiometric timing baselines have been proven to create huge error. This in no way changes the relative timeline one can enjoy using the nuclear decay timing methods, so if you need that security blanket of timing analysis, I'm not going to try to fix on that right now. I will put something in the conclusion section to show you why this somewhat shortened timeline is more accurate.

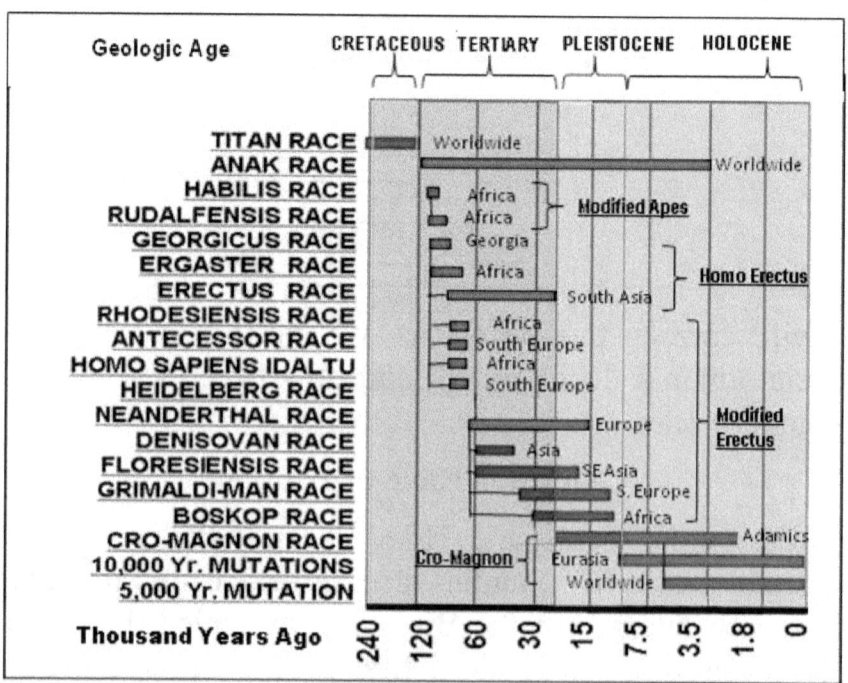

On the chart it is the Homo-Ergaster that became the first African man and woman as described by Haplotype as the Common Ancestors. The Haplotype mutations will give us a pretty good understanding of Homo-Ergaster and Erectus and it is this same science that shows Homo-Neanderthalis [mostly found in mid and northern Europe] had very little contact with Homo-Ergaster or Cro-Magnon man who lived with the Anak people during the Pleistocene as described in the generalized chart previously shown. If you are wondering what the DNA look like, the next image is of live DNA next to dead ones.

Dead DNA- The image to the left is of the 46 DNA Strands found in a typical human cell nucleus. To the right are the same set of DNA strands after a person is "dead". What is the difference? If you give, I will tell you the answer---- "There is NO difference in the carnal existence we perceive". Dead and living DNA have the same sugars, same, interfacing bonds, same structure same chemical transfers, same biology. Don't worry I'm not getting into this area as it becomes strange very quickly.

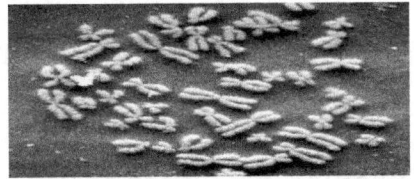 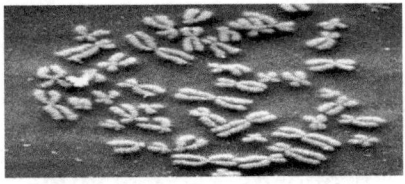

Here are a few of the many things being found recently. Before you read it, let me tell you up front it is a mess.

- Homo-Georgicus [Georgia] have a more apelike skeletal structure than Homo-Ergaster [Africa] or Homo-Erectus [Southeast Asia] suggesting they were the earliest form of Homo-Erectus. This also suggests Erectus came from the Middle East and migrated to both South Asia and Africa.

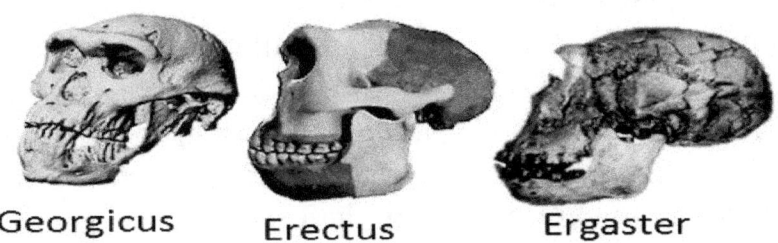

- It also suggests there was not major change in the various "species" that originally were you to express evolutionary changes.

- Even for those who "segregate minor changes, it has been estimated that during 70% of the time period the Homo-Erectus was extant, he was largely unchanged using Nuclear decay dating methods.
- Homo-Neanderthalis DNA has about 1% similarity to northern and East Africans and about <u>0% similarity to sub-Sahara Africans</u>. This seems to show Neanderthal migration from Europe to Africa was rather late.
- Homo-Neanderthalis DNA has about 1½ % similarity to Denisovan. This seems to show the 2 humans have almost no connection.
- Homo-Neanderthalis DNA is only about 2% similar to Asians.
- Homo-Neanderthalis DNA is much less than 1% similar to X Haplotype of the Americans.
- Homo-Neanderthalis DNA is only about 4% similar to Europeans. The Neanderthal people, who lived in Europe for nearly 20,000 years, are not the ancestors of modern Europeans.
- Homo-Neanderthalis DNA is about 5% similarity to Southeast Asian populations.
- Homo-Heidelberg skeletal remains in Africa appear to be more ancient than those in Europe.
- Australian aborigines have the highest % DNA similarity to Homo-Neanderthalis at about 8%.
- Denisovan DNA is about 3% similar to Southeast Asians but only 1.7% similar to Homo-Neanderthalis.
- Mainland Asian and Native American populations had only a 0.2% Denisovan contribution, albeit twenty-five-fold lower than Oceania populations.
- Dental differences show Homo-Neanderthalis has no hominid species common ancestor
- In 2015 it was found that Heidelberg man mtDNA did not resemble that of a Neanderthal. Instead, it more closely

matched the DNA of a Denisovan. The Nuclear DNA more closely resembled Homo-Neanderthalis than Denisovan.
- Over 86 percent of the harmful single nucleotide mutations arose between 5 and 11 thousand years ago.
- Oddly, since then there have been almost no single nucleotide mutations.
- From DNA mutation analysis researchers now indicated that about 81% of the single-nucleotide variants in the European sampled and ONLY 58% in the African DNA sampled arose in the past 5,000 years. This shows Africans have a much older DNA than non-Africans and the 2 groups had very little crossbreeding until well after the 11 thousand year old massive mutation.
- While most DNA scientists have limited their studies to the 10% of our DNA responsible for building proteins. the 90% of DNA considered to be "junk" are now turning out to be more interesting. One thing noted is that they have found 145 'alien' genes not traceable back to our "ancestors". Some believe they came from the Anak. :
- Cro-Magnon appears to have first been in the Middle East and Migrated to Europe and to Asia well before venturing into Africa.
- Later we will see Chimpanzee is has a closer match DNA structure to humans than any other Ape.
- Chimpanzee and Bonobo apes have about 1% difference in DNA
- Studies show Chimpanzee diverged from humans only about 6 thousand years ago and Bonobo may have been even later.
- Earth tilt data suggests the reason Homo-Erectus covered almost all of Africa and Southeast Asia was that they were all on temperate zones. [See following graphic of present, right image, and Pleistocene Earth tilt, left].

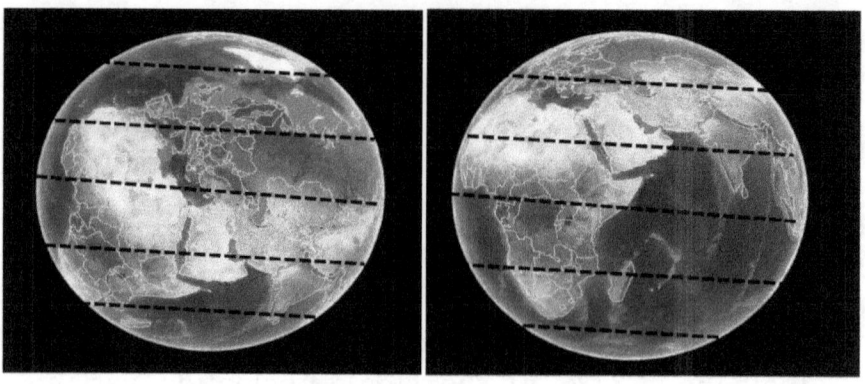

- New details shows that the gait of Homo-Erectus was completely different than Homo-Habilis ape-man in that his feet remained forward while walking. [See following image of Erectus footprint found in 2009, in Kenya. The continued efforts since the initial discovery revealed 97 tracks from 5 sites with at least 20 different individuals, all thought to be Homo Erectus walking like modern man.

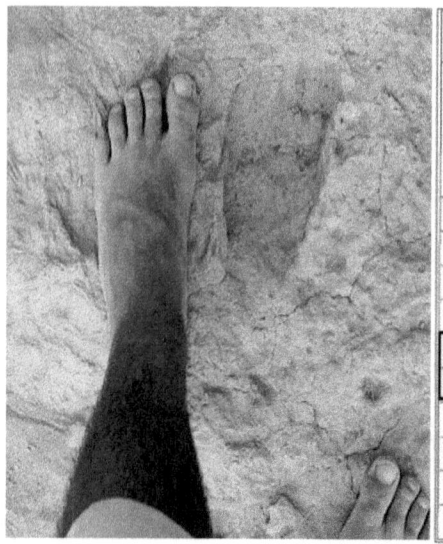

	Base Composition of DNA			
	Percentage of base in organism's DNA			
Organism	adenine (%)	guanine (%)	cytosine (%)	thymine (%)
Maize	26.8	22.8	23.2	27.2
Octopus	33.2	17.6	17.6	31.6
Chicken	28.0	22.0	21.6	28.4
Rat	28.6	21.4	20.5	28.4
Human	29.3	20.7	20.0	30.0
Grasshopper	29.3	20.5	20.7	29.3
Sea urchin	32.8	17.7	17.3	32.1
Wheat	27.3	22.7	22.8	27.1
Yeast	31.3	18.7	17.1	32.9
E. coli	24.7	26.0	25.7	23.6

- Possibly one can tell differences in various animals simply by looking at the mixtures of the 4 ingredients that make up DNA. Please notice from the chart [preceding right] humans are very similar to grasshoppers and more substantially different than a Chicken or Rat. Oops! We

will have to look for a better way, unless you have an exoskeleton.
- The scientists also discovered that the <u>Denisovan interbred with an unknown human lineage</u>, getting as much as 2.7 to 5.8% of their genomes from it. This will make more sense as we go along.
- There is no DNA evidence of the early Homo-Erectus prior to the development of Denisovan and Heidelberg except for a tiny chunk from a group called Archaics.

Homo-Erectus, Neanderthal, Denisovan, and Cro-Magnon humans simply appeared one day- As one can determine from the initial details already presented there is very little probability that these substantially different individuals evolved without using the previous human DNA unless some outside force manipulated the entire DNA strings all at once. In this way 98% of the Denisovan DNA changed from Neanderthal and 98% of the Cro-Magnon DNA could be changed from Neanderthal as well and also have almost no similarity in the Denisovan DNA. To make this even stranger, when this happened, there were enough replicated Denisovan and Cro-Magnon beings to all for procreation. While we do not have enough Erectus DNA to support reasonable DNA studies, we can tell that whatever happened between Habilis and Erectus was huge as there are so many physical differences we can see Habilis was another Ape and Erectus had become a man. As no viable DNA has been taken from the ancient Homo Erectus some of this is going to have to come down to guesswork. Before getting into Erectus, let me very briefly explain the new timing standards for geologic time since nuclear decay has turned out to be so very unreliable and almost always presenting false readings that are much more ancient than known date.

Dating Problems

I'm not getting into this in any great detail in this book, but it is now absolutely known nuclear decay dating shows huge, I mean really huge errors. On December 13, 2006, a magnificent solar flare flung radiation and solar particles toward Earth. Measuring the decay rate of manganese-54 during the flare proved to be very interesting as the decay rate changed hugely during the time of the radiation fallout. It was determined that solar neutrinos zipped through space and affected Mn-54's decay rates used in the experiment. Just think about this. They were testing a single solar flare event and the change was very significant and I'm talking about changing ALL nuclear decay which proved Electron Spin Dating and Uranium Dating, Thorium Protactinium Dating, Oxygen Sediment Dating, Lead-lead-lead Dating, and Argon Dating [which we originally used to date the ages of the Earth] are terribly flawed. The old standard carbon 14 dating also was in jeopardy beyond 30 thousand years which limited the number of flare events or cosmic ray events, or high temperature events and all the other things that either added or removed protonic components. It was also found that the decay rates of silicon-32 and radium-226 even showed seasonal variation, according to data collected at Brookhaven National Laboratory on Long Island and the Federal Physical and Technical Institute in Germany. This error was just the material sitting there with almost no outside interference.

Wood buried in igneous rock in Queensland Australia has been dated to 40 thousand years, while the basalt around it dated to 45 million years.

Both dating subjects should have given the same date, since the igneous rock was formed at the same time the wood was buried. Many of the "data-ologists" don't tell you about major errors like this.

Excess argon-36 was found in three out of 26 lava flows in recent times. So Argon/argon testing would show a much older date that actually was "KNOWN". This is believed to be because there was too much of the argon-36 in the first place.

In the Grand Canyon "lava flow testing" showed <u>lower levels of lava were younger than the top layers.</u> At different volcano sites, that had eruption in 1949, 1954, and 1975. The same thing was noted. Geochron Laboratories of Cambridge, Massachusetts dated these samples.

Even though the oldest of these samples are just over sixty-years old, the lab tests provided ages that ranged from 270,000 years to 3.5 million years old.

Additionally, we go to Mt. St. Helens and its eruptions in the 1980's. Samples there gave <u>old ages in the range of 300,000 to 2.7 million years.</u> Hopefully, you are beginning to see that we know less about how old we are than you believed before reading this. If neutrinos from a single solar flare can make things look older, what if the entire Earth was closer or farther from the sun? I know that sounds odd, so just keep it in the back of your mind right now as we try to find some standard for dating. Nuclear timing fails miserably if there was ever any nuclear events, high sunspot actions, any high temperature events, and any cosmic ray intrusions almost always provides a date that much older. As a confirmation of nuclear decay timing, something called strata positioning is used. That is where something is found across "strata lines that correspond to different nuclear timed ages. Unfortunately, it has been found that depth of a bone is almost always inappropriately determined as the earth shift incidents continuously change strata positioning. The following images show how it appears that trees stood for millions of years as they passed through many rock layers identified as separations by time.

There are many, many books written about these two so often used timing methods, but for the sake of brevity, let me just introduce you to the new timing.

You will find a difference in timing between my charts and those you normally see and you think the data is all wrong simply because you still are inundated with the known to be wrong timing methods. I'm not going into why a scientist or teacher would want to mislead someone when many already know that we are not saddled with not having a standard for timing geologic events. In fact, we have 45 timing standards to assure proper cross comparison of the last 5 extinction periods and we can simply use relative timing principles to match the reference marks to a more reasonable timing using the new dates for the horrible Permian Extinction [480 thousand years ago], Triassic Extinction [340,000 years ago], the Jurassic Extinction [240 thousand years ago], the Cretaceous Extinction [120 thousand years ago] and the Pleistocene Extinction [10 thousand years ago]. All these events are recorded in Ice Core thermal and CO_2 data [describing intricate variation in respiration and atmospheric temperatures], Magnetic shift data known a Paleo-magnetic tracking [tracking Earth magnetic field changes across time], Island Hot Spot tracking of Hawaii [tracking modifications in the Earth axis of rotation], and O_{18} isotope [calcium shell fish density tracking]. When taken together these timing metrics are both stable and comparatively assured. The details of these timing methods are not the purpose of this book, but many have not been introduced to the errors of nuclear decay. The following timing graph describes the 4 timing methods of the last 4 extinctions so there is less confusion and I will provide some details showing the old timing method as we go along as well.

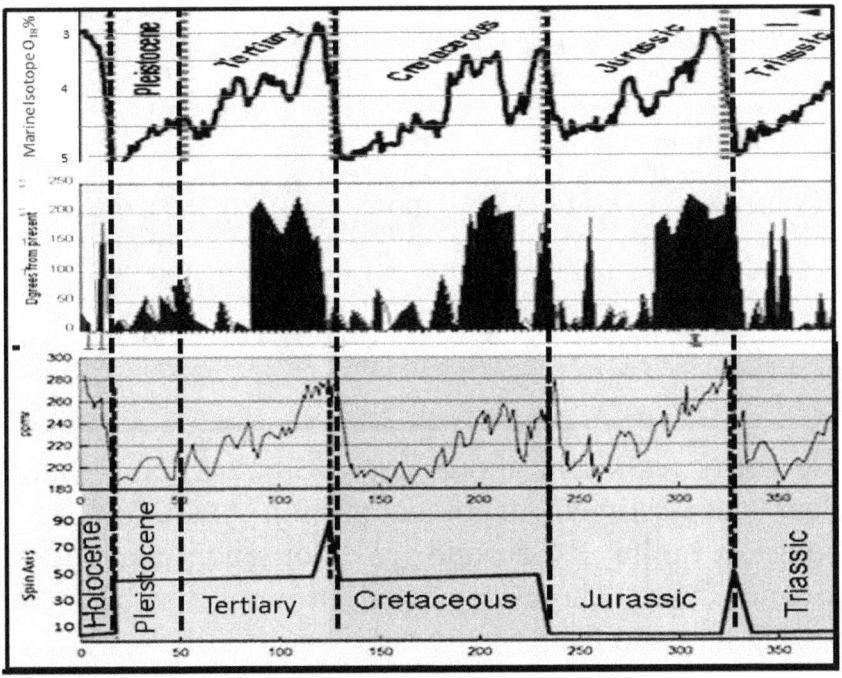

To make this new timing even more reasonable, the characteristics also describe how the Earth shifted during each extinction event. The one we care about in this book is the 30 degree axis shift that occurred 10 thousand years ago which has been confirmed by the almost immediate "million individual Mammoth herd extinction" as Siberia shifted into the Arctic circle quick freezing the animals with flowers still in their mouths.

The following geologic timing chart shows the timing previously presented as "nuclear decay based timing [that you were taught in school] and the more modern characterization of CO_2 density, Temperature, O_{18} levels, Paleo-magnetic and Hot spot travel timing. The farther from the 40 thousand year limit the more variable the nuclear decay timing gets but the other timing method have no such limitation.

Standard Geological Timeline

Era/Period/Epoch	Time (M yrs. ago)	Time (T yrs. ago)
Archaeozoic Period	5000-1500	50,000-3000
Proterozoic Period	1500-545	3000-1000
Cambrian period	550-500	1000-900
Ordovician period	500-440	900-800
Silurian period	440-410	800-700
Devonian period	410-365	700-600
Carboniferous	365-300	600-500
Permian period	300-250	500-400
Triassic period	250-212	400-300
Jurassic period	212-145	300-200
Cretaceous period	145-65	200-100
Tertiary period	65-0.04	100-40
Pleistocene period	0.04-0.01	40-10
Holocene period	0.01-0	10-00

While many texts talk about Homo-Erectus being found and nuclear decay timed to be 2 million years old, a more reasonable timing is around 75 thousand years ago. The following table shows updated timing for the various emerging humanoids.

Million Yrs. Ago	Thousand Yrs. Ago	Major Humanoid Characteristics
100	200	Titans & Dinosaurs
65	120	Anak & Mammals
32	115	Horses
16	107	Great Apes
8	100	Australopithecus
4	92	Homo-Habilis
2	84	Homo-Georgicus
1	76	Homo-Ergaster
0.5	68	Homo-Erectus
0.2	60	Heidelberg
0.1	50	Neanderthalis
0.04	40	Cro Magnon
0.01	10	First Mutation
0.05	5	Second Mutation

I say all this to set the stage with the first humans on the planet called Homo-Gigantus. Some simply call them the Titans. Many have tried to ignore this important group and have introduced so many unresolved issues by the exclusion that it is certainly criminal. As I want to stay on your good side, I'm going to reintroduce them. Even if a teacher introduced the term Titan to you it was probably to discount the Greek historical records as mythology. Today we are finding out that, while the stories have been glamorized, there is a lot of history provided as well that is certified by many pieces of evidence and our Bible.

Gigantus to Modern Man

The first humans are sometimes known as **Homo Gigantus**, but, like I said you may have heard about them as the Titan people who became the Anak people who ruled over "normal men for thousands of years. Not only did they rule over "other humans" they pushed the belief that they were gods rather than men. In the book of Genesis, chapter 6, we find that this ancient society of giant people lived well before the time of Adam and in chapter one; Moses tells us that our Creator God made a new man to <u>replenish</u> the world after the Homo-Gigantus people had died off.

Genesis 1:27-31- So God created man in his own image, in the image of God created he him; male and female created he them. And God blessed them, and God said unto them, Be fruitful, and multiply, <u>and REPENISH the earth,</u>-- And the evening and the morning were the sixth Yawm' [Age].

Bible tells of 4 separate "developments of Man, but that is a tiny drop of information concerning races of mankind.

Genesis 6:2-The "Giants of Old" or Homo-Gigantus [They lived before the time of Void or extinction. According to Judeo-Christian documents, this group became the angels described thought the Bible]

Genesis 6-3-The *"Anak"* people [This group came from the Angels, so they were really the same Giants, sort of, re-established]

Genesis 1:30-The *"humans of the 6th Age"* or Homo-Erectus after the Void

Genesis 2:3-The *"Adamics"* or Cro-Magnon during the 8th Age

The first race described as ancient humans, Giants of Old or TITANS that walked with dinosaurs. Some call this group Homo Gigantus. They lived; built cities; increased high technologies; and had wars during the Mesozoic Era. Some believe that when they died, these Titans became the entities we call watchers or [angels] identified in Egyptian, Christian, and Jewish histories, but in this book that doesn't really affect things. The Cretaceous Period ended with a bang and the Bible indicated the Earth had no form and was void as all the cities had disappeared. While there was a massive meteor hitting the Yucatan that helped the end come to the Age of dinosaurs, a group of the Homo-Gigantus now called the Anak [Anak is Hebrew for giant with a long neck or head.]. Interestingly, we have uncovered a number of skulls around the world in Peru, United States, France, Egypt, Russia and other parts of the world that are Dolichocephalic [long and very thin]. Following are a few of over 300 that were found in Paracas Peru. To be clear, the remains shown are not of the original Anak who lived before the Pleistocene Extinction. They were the descendants of this first group. It should be known that this group has not been adequately tracked and tested by DNA sampling and Haplotyping.

Unbelievably, DNA samples from this group show some type of "Alien" component. Anyway! Almost all the dinosaurs died off and the Anak people survived during some difficult time. Let's look at books of Jeremiah and Isaiah in the Bible that sort of expand on the Genesis indication that the Earth became void and without form.

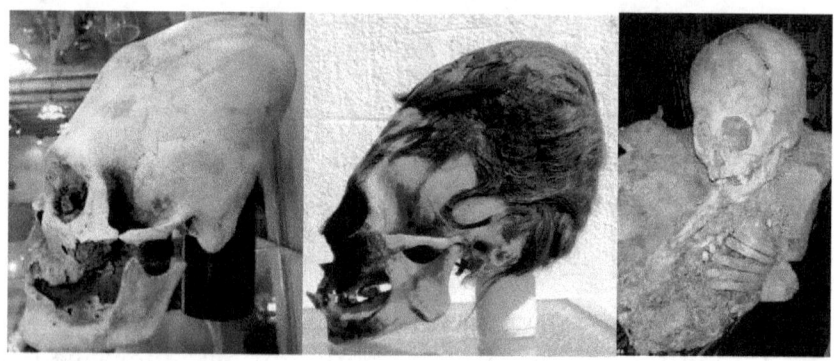

Jeremiah 4:23-24-[near the end of the Cretaceous Extinction] *"I beheld the Earth, and, -- all the cities [Titan cities] thereof were broken down.* [If the cities were broken down, there must have been cities before the war.]

Isaiah 9:17-21---Is this the man [Satan leader of the Anak people] *who made the world like a desert and overthrew its cities [Titan cities], All the kings [Titan Kings] of the nations lie in glory, each in his own tomb.* [Kings of nations and Cities shows a substantial civilization.]

"Revelation of Adam and Seth" *[Jewish Gnostic Text]-Then the earth trembled and the towns were shaken; the birds fed to satiety upon the dead. The earth became [void] and the universe became a desert.*[The first heaven war destroyed many cities that had previously covered the earth.]

"Origins of the World" [Jewish Gnostic Text]--*Heaven and Earth were destroyed by the troublemaker that was below them all. And the six heavens shook violently; for the forces of chaos knew who it was that had destroyed the heaven that was below them. And when God's Spirit knew about the breakage resulting from the disturbance, she sent forth her breath and bound him and cast him down into Tartaros* [After the first of the heaven wars, the Earth became void.]

After the destruction, those responsible became a somewhat different type of "Human". They mutated into what is commonly referred to as the Anak Race. The Anak looked like us, but we are told they had their "Light/Spirit" taken from them. Here are a couple of the many Judeo-Christian texts identifying this "punishment".

Jubilees 2:9-*Nor may we take revenge on him because he has stripped us of the "light". He marked out the borders of the world and created man in his own image with whom he hopes again to people heaven, with pure souls.*

Book of John the Evangelist- *"My father [God] changed his [Satan's] appearance because of his pride and the "light" was taken from him. His face became like a heated iron wholly like that of a man".* [The old "heated iron man head" thing.]

Seriously, this light or spirit has been are really bad problem for the Anak as they could never leave this universe and go to the universe known as Heaven. Unfortunately they still died, but once dead they became something called demons. None of this has anything to do with this book, so let me stop this discussion and describe the Anak while they were alive, because they were the same as you and me, just big and they lived a long time.

That being said, they still looked like the Titans they originally had been. They were still giants and possibly had the long heads, but after this time the Bible called this group the Nephilim or Anak people. You can imagine it was hard on the Anak at first. Civilization had vanished and technology had been destroyed but the geneticists of that time took the remaining animal populations and began modifying DNA. There are very many books about this it. In Sumerian, Greek, Egyptian, and Judeo-Christian histories that are supported by the massive rise and unbelievable diversity in animal populations as the Tertiary Period took hold. We are told God didn't like it at all and called the animals modified by the Anak "UNCLEAN" or abominations. The Anak people even reinstated the Dinosaurs later on and today scientists are finding multitudes of dinosaur remains that are <u>not fossilized</u> meaning they were, again, living during the Pleistocene Age. The Anak began experimenting on Apes as well. But each try simply made another type of ape. The more successful ape-man varieties were called Australopithecus and Homo-Habilis, but they would be little better than pets to the Anak. The Anak left the Homo-Habilis in Africa. I know you haven't been taught this as teachers want to remove and possibility of slipping up and saying the God word, but the physical evidence continues to drive away uncontrolled evolution as a cause for any major expansion of capability as the Law of Entropy always drives variations to a higher state of DISORDER rather than advancement. We will look as some of the massive amounts of evidence in a bit, but mostly we will look at this next "phase" in human development as the Ape-man, Homo-Habilis, seemed to have a massive conversion in capability, size, method of travel, and ability to serve the Anak people. It was as if there was a brand new creation. [As a note, toward the end of this book we will talk about another "Apish-Man" group called the Vanara People in ancient tests. These looked like Homo-Habilis, but they were integrated in society between 3500 and 2500 BC. Never mind! Forget I said anything. I want it to be a surprise.]

Homo-Erectus Man

The Sumerian texts tell us that the Annunaki [same as the Anak people] cried out to God for a servant to help them live better. Possibly some biologist wanted to continue modifying apes, but most knew a major change was required. According to Judeo-Christian texts, this "new Man" appeared during the 6th Age after the Earth became Void [Cretaceous extinction].

We call this guy the Homo-Erectus. During the Tertiary Period Homo-Erectus changed quite a bit, but most references indicate that all animals were "manipulated" such that most were called "unclean abominations" during the Tertiary and into the Pleistocene Age. Sumerian, Chinese, Jewish, and many other historical references attest to some type of genetic manipulation that was not always successful, just like we do today.

Here is when the history gets strange as the Neanderthal Race, and many other races came from Homo-Erectus, but we are told they were not created by God. Instead, the ANAK tried to make Homo-Erectus a better servant, according to many texts. The outcomes were the Neanderthal, Heidelberg, Denisovan, Antecessor, Peking, and Rhodesiensis Races as well as others we will never know about. Today, we find that there are "alien genes" in Neanderthal man and in the Paracas Race, but most simply ignore the implications.

Cro Magnon Man

Finally, Cro-Magnon Man or Adamic man was created during the 8th Age after the Earth became void which we typically call the beginning of the Pleistocene Age. 40 thousand years ago. As the Pleistocene Age ended, 10 thousand years ago, massive mutations caused substantial change in humans and many racial traits were formed. We know this happened by detailed Haplotype DNA mutation tracking which has changed the way we view history. The Titan Race started it all. Homo-Erectus began changing, mostly by outside forces and then Cro Magnon came along. The Homo-Erectus people stayed more to themselves but after years they would finally integrate and be modified to form a number of different types of Erectus based humans. After Cro Magnon came they mixed with the various types of individuals still extant. If you remember the previous chart on DNA differences [shown again –next], no one has really done a review of these two types of modern humans so that we can gain more insight and I believe it an uncomfortable detail to disseminate as one type of human would claim superiority of DNA over the other. to investigate this known difference.

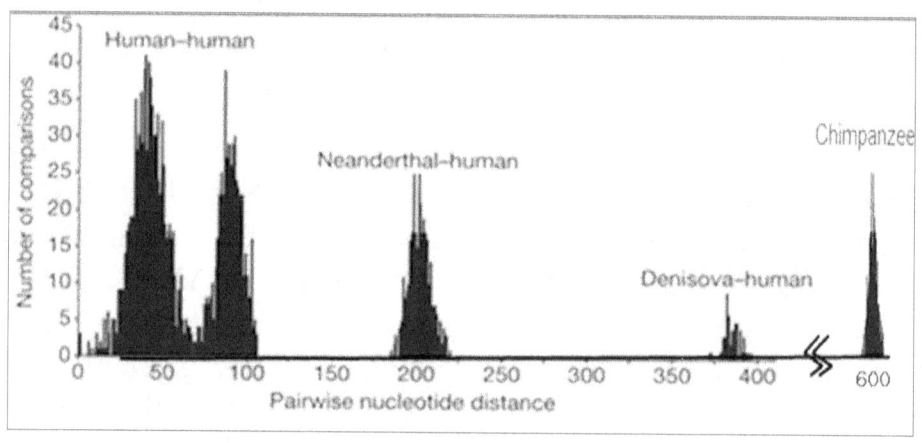

Chimpanzee and Giganticus

For completeness, I have a chapter at the end of the book describing the strange appearance of the Chimpanzee. It truly has little to do with the Homo-Erectus, but I could not help myself. After this quick overview, let me get back to the Homo-Gigantus so you won't think I'm telling you a fairy tale. By the way, when I talk about Homo-Gigantus I don't mean "Giganticus". That creature was a huge ape-man that looked like the Sasquatch that is reported today. The following image shows what we believe a 9 foot high Giganticus looked like and an 11 foot Anak [or Gigantus] next to a modern man and Chimp. Gigantus looked like us only much, much larger. As large as he was, he still had to watch out for dinosaurs.

Of interest, Goliath and King Og from the Bible history were 11 foot tall, so they would have looks like the first guy. It's no wonder that when the Jews saw the Anak living in Canaan they said they felt like little grasshoppers.

Homo-Gigantus Beginnings

While not much is known about Titans simply because they were here so long ago, there has been physical evidence and written texts that help us understand about this first important race of man. Most of us have read about the Titans in the Greek histories, so I'm not going to get into that, but let's look at some of the finds around the world showing how this race of people lived around the world during the time of the dinosaurs.

Ecuador: Tools so ancient we cannot date them were found in Ecuador. The main thing that is noticed is that they could only be used by a massive race.

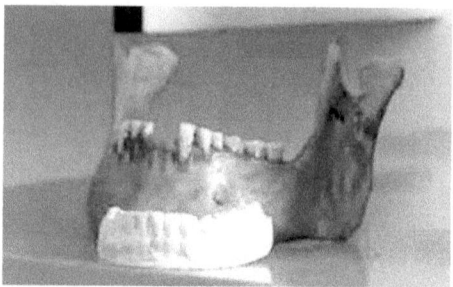

Around the World: Similar massive tools were found in Malta, Australia, and around the world.

United States: While many might not know it, hundreds of giant skeletal remains have been uncovered in the United States. Here is the important part here. Some of these skeletons turned black when air hit them and they became powdered because they were so very ancient. One massive skull was found to be petrified. The size of the jaw was easily determined to be strange as shown above right. The "normal jaw is shown in front of the "jaw evidence" of the Titan race in America.

Texas Feet: One thing is for sure if Titans lived here, they would have big feet and possibly even be found with Dinosaur feet; and so it is. The footprints are massive and show a race of 10 to 12 foot tall Titans lived during this time. Note the one of the left going the opposite direction as the Dinosaur as they shared the beach.

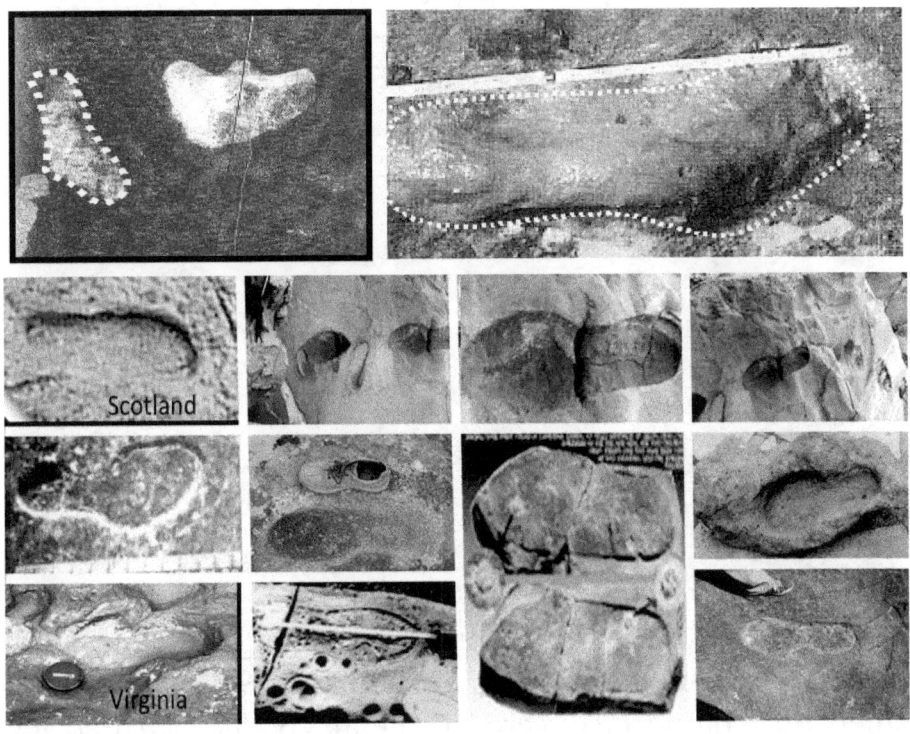

Dozens of 18 inches long human footprints were found at Glen Ross, Texas and around the world. Some of the footprints and shoeprints are found alongside dinosaur prints. The prints have been studies for years and conclusions force the reality that these giant humans lived during the time of the giant dinosaurs.

Many: I know you are thinking a couple of footprints can't be used to form an entire race of people, but what about hundreds? The more they look the more they find. Hundreds of human footprints alongside, dodging the dinosaurs are being found around the country and the world. A small sampling of the many spots are shown below.

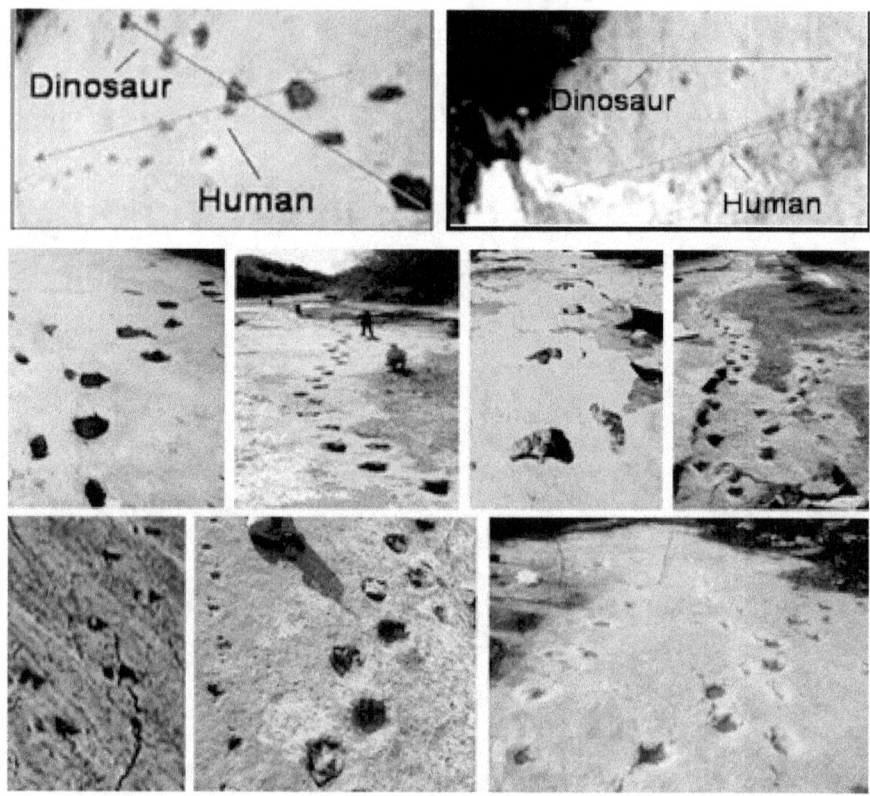

Stone Bone: If these guys were here so long ago some of the skeletons would have been petrified. In 1982 a Researcher named Ed Conrad discovered in Pennsylvania petrified teeth inside the jaw-like area of solid rock. Then he found more and more specimens that bore the contour of human bone. You can just call it a stone head, but petrified skulls tell a tale.

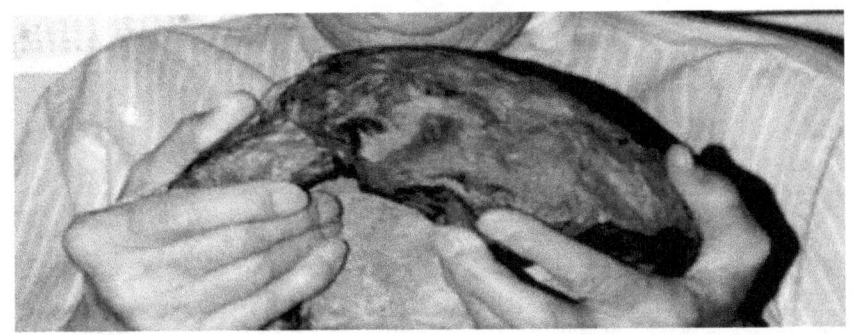

Wilton M. Krogman, the internationally acclaimed bone expert identified the first "stone bone" as a human calvarium, a portion of a skull with the eye-sockets broken off. A year later Ed Conrad discovered the large boulder in which was embedded the object that bore a distinct resemblance to a huge human cranium.

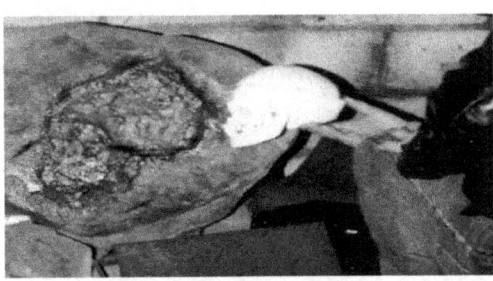 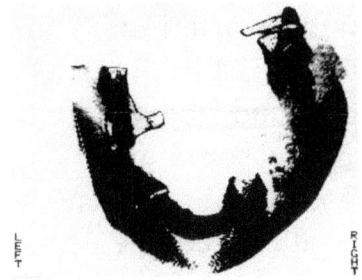

A CAT scan had been done of this particular specimen and revealed intriguing characteristics of a human skull jaw and joints. [See above right]

Turkey: In the late 1950's during road construction in Homs southeast Turkey, many tombs of Giants were reportedly unearthed. These tombs were 4 meters long. During exhumation, the skeletal remains were examined. The human thigh bones were measured to be 47.24 inches in length. They calculated that the person who owned this Femur probably stood at fourteen to sixteen feet tall as shown to the right. A cast of this bone is shown below. Images of similar sized giants were shown in the United States, see below right.

 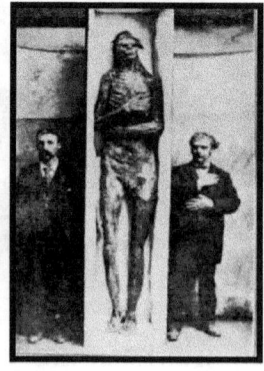 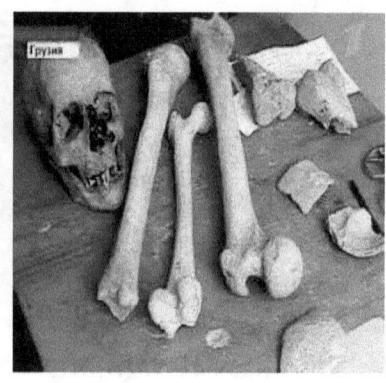

Russian Giants- Around the world we find these Homo-Gigantus bones. The picture above right shows a normal leg bone next to one of the giants found in Russia. The middle image is from Ireland, possibly one of the Fomorian giants. He has been petrified and it was indicated he had 6 toes.

Australian Titans: In Cosmo Newberry, In July 1970, an extensive, 4.5 kilometer long trail of giant-six toed footprints was found about 540 Kilometers north-west of Perth. Each print measured over 16 inches long, displaying a soft pad and opposable big toe. The man's size was estimated to be over <u>11 feet tall.</u> A couple of the footprints are shown next. [It should be noted that some of the massive footprints in Australia have been attributed to the Homo-Giganticus [Ape-man] rather than a Titan.

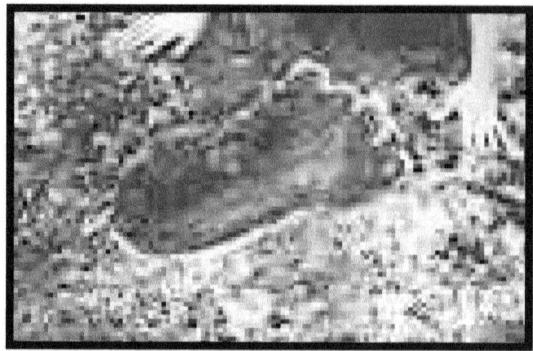 

Technology: Besides bones and feet, we have found a large amount of physical evidence of high social order and science of the Titans. Almost impossible to fathom, we have found shoeprints, batteries, nuclear power, brass manufactured good, and many other extremely ancient components of this once great society. After the war and the Cretaceous Extinction rocked the world, the Titans had vanished as did almost all the dinosaurs. On this barren Earth came a new race of men called the Anak.

Before we get into the strange evolution of ape into homo-erectus, let's see if there is information and evidence of a giant race of "modern looking" men, with red skin and hair, 6 fingers and 6 toes, double rows of teeth, and a long thin Dolichocephalic skull that lived with the dinosaurs. Genesis 6 simply called them the giants of the past,

Mayan Creation- *"Popul Vuh"-In the beginning was total silence. God created the world and the giant gods. This first race [of giant men] was capable of all knowledge. They examined the four corners of the horizon; the four cardinal points of the Firmament; and the Round surface of the Earth. There were many conflicts between the gods.*

Peruvian Creation-*Inca -Viracocha created the world. Initially it was dark. Giants [Giant humans]were made from stone and ruled the world. The giants ignored the creator's wishes and did not worship him.*

Greek Creation-*Long after the "beginning of time", the void known as Chaos came into existence in the universe. 12 Children of Uranus [sky] & Gaea [Earth] became the first Titans [Cronus was the bravest]A second Generation of 8 Titans included Atlas, who held the sky, and Prometheus, who helped man in opposition to Zeus]* [The ancient men became angels. I don't know if only 8 were transformed, but that was the Greek belief.] *Zeus and the other 4th generation gods came from Cronus.* [This was talking about the conversion of Titan/Angels to the Anak people who thought themselves to be like gods.]

look at some of the evidence of the Anak. Our first stop is the United States. Thousands of the skeletal remains of giants that inhabited North America in the distant past have been found in just about every State. By far the Ohio Valley with its massive copper mines was the most populated. Let me give you a general view of these people. They were red-head, white people, with 6 fingers on each hand, and they had long massive skulls, some of which would fit over a "normal person's head. Some of the bones so old they crumbled at the touch and others fairly well preserved and originally many had stood over 12 feet tall. They are believed to be well over 20 thousand years old to crumble when exposed. Over 1,500 accounts from newspapers and books published in the 1800s and early 1900s describe how the United States area alone had been filled with people between 7 and 18 feet tall.

Rather than pulling in a large amount of newspaper, scientific, and historical records, here is a pretty good accounting of the giant Anak remains found.

- Ohio - 3000 Anak skeletons found
- Arizona- 53 Anak skeletons found-a number were <u>twelve feet tall</u>
- Minnesota- 36 Anak skeletons found
- New York- 35 Anak skeletons found
- Pennsylvania- 30 Anak skeletons found. One, found 12 feet deep was described as <u>18 feet long</u>, had a 9 foot sword, an almost rusted away massive helmet, and a double row of teeth.
- Georgia- 3 Anak skeletons found
- West Virginia- 8 Anak skeletons found
- Illinois- 14 Anak skeletons found
- Nebraska- 12 Anak skeletons found
- Tennessee- 11 Anak skeletons found
- Indiana- 9 Anak skeletons found
- Nevada- 8 Anak skeletons found - <u>red-haired giants</u>

- Alaska- 7 Anak skeletons -massive skulls was sent back to the Smithsonian for study that were about 70% more massive than current humans
- Iowa- 7 Anak skeletons found
- Kentucky- 3 Anak skeletons found; one 12 feet long.
- Missouri- 1 Anak skeleton found 12 feet long.
- Colorado- 3 Anak skeletons found
- Hawaii- 3 Anak skeletons found
- Montana- 2 Anak skeletons found
- Wyoming- 1 Anak skeleton found
- Mississippi- 1 Anak skeleton found
- Utah- 1 Anak skeleton found
- Florida- 1 Anak skeleton found, it was so ancient that once it was disturbed, much of it simple became like dust according to scientific record.
- California- 9 Anak skeletons found. One was 12 feet long. A number of the remains are shown in the graphic below.
- Wisconsin- 14 Anak skeletons found [Next left shows the long skinny head. One was 12 feet long another was estimated to be 14 feet long.]
- New Mexico- Anak skeletons found [See middle below. The small jaw is a "normal" size.]

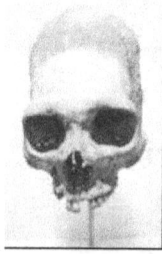 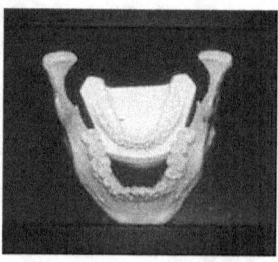 

- Arkansas- 8 Anak Skeletons found up to 10 feet tall, the last in 1976, several shown below left.
- Texas- 3 Anak skeletons. One skull shown in graphic compared to normal skull.

In the Ohio Valley, the ancient Adena people buried these giants under massive human filled hills as shown below.

They called these giant overlords the **Ronnongwetowanca**, but I like Anak better. These same overlords of the Adena were also featured in the Ojibwa chronicles but were called the Archaics. It is estimated they finally overthrew these giants around 3000 years ago; however, some of the giant remains listed above were dated to be as recent as 2000 years ago.

Adena Story- As I mentioned before, the Adena were a Cro-Magnon group who first settled in the Ohio Valley, but spread to about 1/3 of the continental area. Some of their spreading was to get away from the Ronnongwetowanca [Anak people] who enslaved them. These Ronnongwetowanca people were also called the Archaics. With thousands of mines dug and mined over the centuries, it is believer the giant Archaics had enslaved a lesser group we call the Adena. After a long time of enslavement and heavy burden, the Adena rose up and destroyed the Archaics. While they continued to mine, they also built giant mounds in their spare time. They also would not survive as they would be destroyed by the Hopewell,

Aztalan and Ojibwa. The Ojibwa only know stories about the ancient Adena.

According to Indian Legend- There *were two different races of strange humans that pre-existed their culture. One was the <u>Archaic people</u> who had slender bodies with long narrow heads. The other group was the <u>Adena people</u> who had a massive bone structure with a short head. The Archaics were living in the Ohio River Valley prior to the Adena. In what is assumed to be around 1000 BC, the Adena moved into the area, coming up from the South, to claim dominion over the land. <u>After a great war</u>, the Archaics were destroyed by this more advanced and powerful race. From the Adena the art of mound building was established.*

David Cusic, a Tuscorora Indian wrote about the end of the Archaics in legend- He wrote-*There was a powerful tribe called Ronnongwetowanca [Archaics]. They were giants, and had a "considerable habitation." When the Great Spirit made the people, some of them became giants. They made themselves feared by attacking when most unexpected. After having endured the outrages of these giants for a great long time, the people banded together to destroy them. With a final force of about 800 warriors, they successfully annihilated the abhorrent Ronnongwetowanca [Archaics]. There were no giants anywhere after this, it was said. This was supposed to have happened around 2,500 winters before Columbus arrived in America.*

The map following was created by researcher Cee Hall showing the sights he found that have been reported.

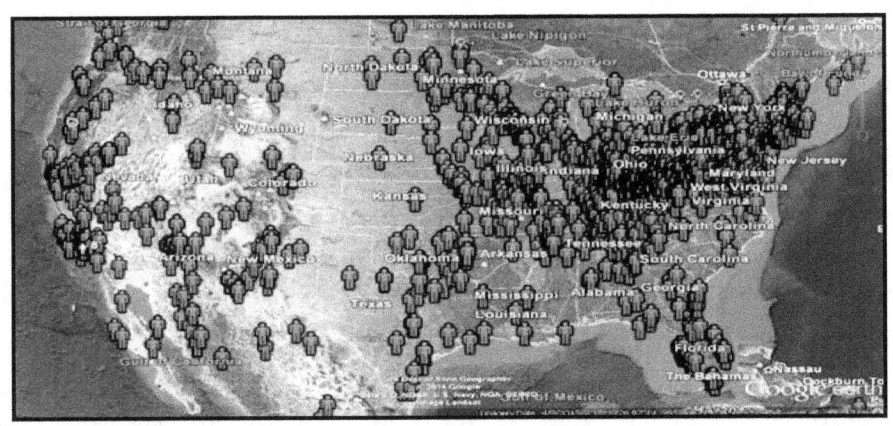

American Archaic Giants On Display

In 1930 Don Dickson discovered what was described at the time as the largest Neolithic burial site in the world, 90 miles south of Peoria, Illinois and he unearthed 248 of the 3000+ skeletons, for an open-air museum. The image following shows a small segment of the museum. *"While some of the skeletons appeared to be only 12 thousand years old which predated the Adena, all appeared to be modern in appearance and many were of extreme age and gigantic in stature."* These gigantic people would have been some of the Anak or their descendents Anakim.

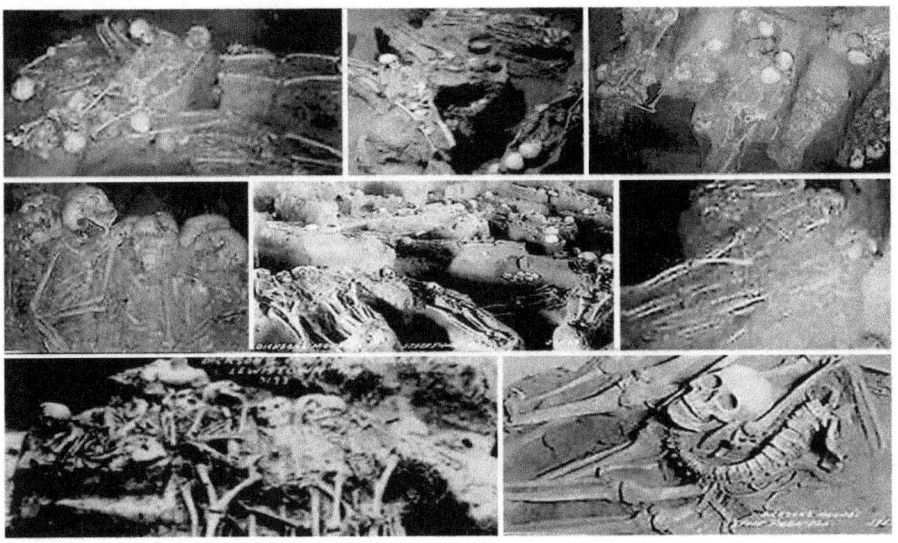

Irish Giant-This photo of a 'fossilized Irish giant' was taken at a London rail depot, and appeared in the December 1895 issue of Strand Magazine. The giant was dug up while prospecting for iron ore in County Antrim, Ireland. The Royal College of Surgeons, made a skeleton using the fossil dimensions. It was over 12 feet high and weighed 2 tons. As with many of these ancient giant people, it showed he had 6 toes on its right foot. After being exhibited in Dublin, it was brought to England and exhibited in Liverpool and Manchester at sixpence a head, 'attracting scientific men as well as gaping sightseers'. This specimen [below left] disappeared in the early 20th century.

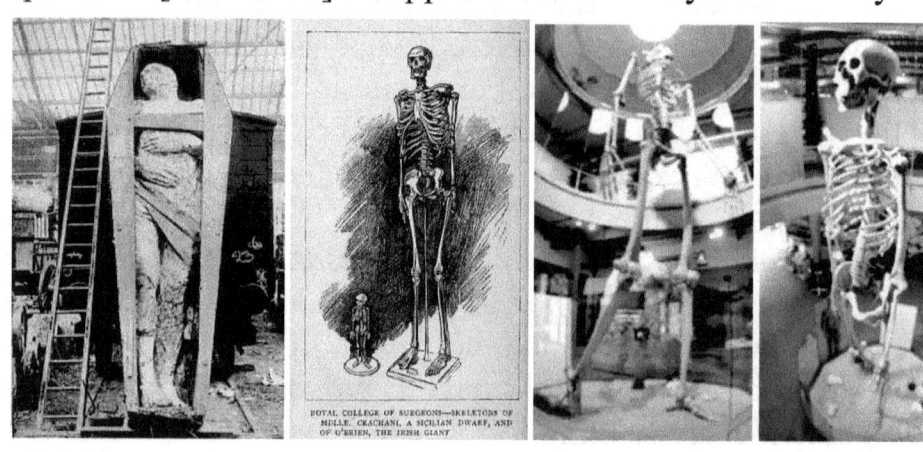

Just one of many skeletons artifacts of the South and Central American Akamim Giants who ruled during the Pleistocene and into the Holocene, this modern looking, 20 foot tall, ancient giant skeleton is on exhibit for world to see in Ecuador. Akamim history was provided by the Ulga Mongulala people of Brazil which even described a massive war that ended the Age of Blood about 3100 BC. [See last 2 images above]

Besides being taller than normal humans, the Anak had been living in a society for many thousands of years and had established a society similar, in some ways to our own. I don't mean by just wearing shoes and walking with dinosaurs. I'm talking about a society with air travel, electricity, nuclear weapons and genetic research. It is the genetic research that interests us the most when understanding the many varies of Homo-Habilis and Homo-Erectus that seemed to come from nowhere while almost no mutations in modern DNA describe the insane adjustments being made in the early Ape-people and humans.

Anak Civilization

Whether you call them the Anak, Anak, Annunaki, Lords of Amenti, Olympians, Arya, Akamim, Giants, Ronnongwetowanca, or even gods or demigods, there is no fantasy in these people living, breathing, ruling, interbreeding, warring, and traveling around the world. The stories around the world are all the same or very similar. If you wondered why Greek gods could have the same limitations and frailties as "other people" such as greed, hatred, envy, lust [lots of lust], laziness, pride, selfishness, bitterness, desire for power and all the rest, it is because these guys were PEOPLE. The stories have been "glorified" just like our heroes [Abraham Lincoln comes to mind] have been almost deified, these rulers of just about the whole world must have been something. If you want to ignore them and ignore hundreds of documents that tell about them and their limitations and ignore the hundreds of pieces of physical evidence, found around the world, be my guest, but don't even try to understand how our DNA has been mutated, distributed, and interbred. You'll get the same mess we are dealing with today with the Haplotype mapping being so mixed up that history, religion, physical evidence, and science cannot agree with what the DNA describes. Yes, the Annunaki people [Sumerian name for these guys] lived a very long time, but so did the "other people" before 6 thousand years ago when a massive mutation of our DNA changed us forever. We generally know what they looked like, some of their behaviors, and their capabilities.

Don't even go and think that EVERY PERSON in the old world was an idiot and described these [Greek god] people to scare children into acting right. That makes no sense at all. Some of the texts talk about ugly looking and dumb Anak; others talk about whinny Anak, others tell about their understanding of genetics that can explain why kangaroos somehow magically were found on Australia after a massive flood that consumed the whole world for a time. The Dogon tribe of Africa learned about stars and planets that have only recently been seen by our most powerful telescopes. They didn't even have a small telescope, but they did have the Anak people visit [They called them the Dogu]. Indian history talks about Mercury engines invented by these guys well before modern man knew that Mercury made the most efficient and powerful engines. In South America, the pre-Inca Anak grew massive boulders together to form irregular stonework that smashed together so tightly that no space can be found between the manufactured blocks. In Russia, the remains of Anak tungsten and titanium micro coils were found. Most of the Siberians don't even know what tungsten is today.

Russian Anak-On the eastern side of the Ural Mountains, pieces of small metallic springs, screws etc. have been found, in the 1990s, where there is no explanation if we look at normal Haplotyping. One of the pieces is shown below.

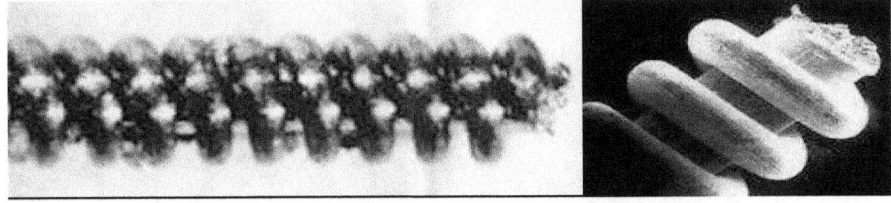

The size of these things ranges from 1 ¼ inches down to extremely small pieces of copper, Tungsten and Molybdenum. Reportedly, these things have been found in the thousands so we are not looking some rare anomaly. Similar exotic metal works have been found in Peru, Central America, and other parts of the world.

Russian Sky View Images- While we are on the ancient Russians, let me show you one odder thing either done by the Anak or for them when they flew overhead. This giant Elk image can only be seen from the air. Discovered very recently, this type of thing has been found around the world. If you can't see the image, it has been highlighted in the image to the right. Similar sky view images have been found in England, Peru, and the United States.

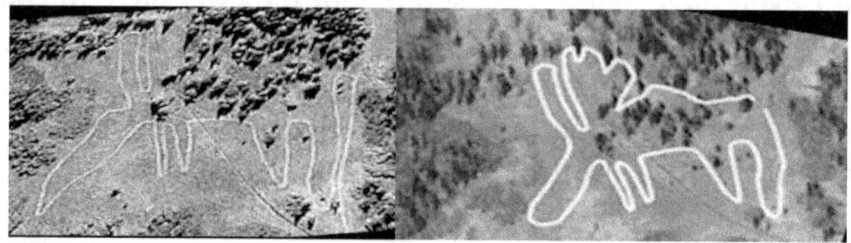

In 1932, local pilot George Palmer was flying over the Colorado River into Blythe when he accidentally spotted the formations below him. Intaglios are gigantic human, animal and geometric figures on the ground surface. There are over 300 intaglios in the American Southwest and adjacent Mexico. The best known of these mysterious figures are the Blythe Intaglios. The ground drawings are situated on two low mesas or terraces. There are several figures in three locations. The figures include two large humans, a feline and a concentric circle and a spiral. The largest human figure measures 171 feet from head to toe.

Tangut Astronomical Insight-This Central Asian tribe, the Tangut, had a strange concept that went along with the apparent Biblical concept of inhabited planets. They believed there were 11 major luminaries including the sun, moon, Mercury, Venus, Mars, Jupiter, Saturn, Tsi-Tsi, Ouebo, Rahu, and Ketu. How could they have known about other planets beyond Saturn? No one knows the answer, but they knew about them just the same. This raises an interesting question that doesn't stop with the Tanguts. How did the ancients know about planets that were too small to see with the naked eye? [The way they probably knew was that some of their Anak ancestors had visited the places or had powerful telescopes.]

6000 BC-Geode Battery- Electricity is a sign of civilization. In New Mexico was found a piece of an electrical connector of some kind <u>embedded in rock</u>. X-Rays showed the pins were made of steel. Besides that, a battery was found <u>inside a California Geode</u> as shown below. I placed a common "D-cell" next to it for comparison. These would have been in used over 8 thousand years ago.

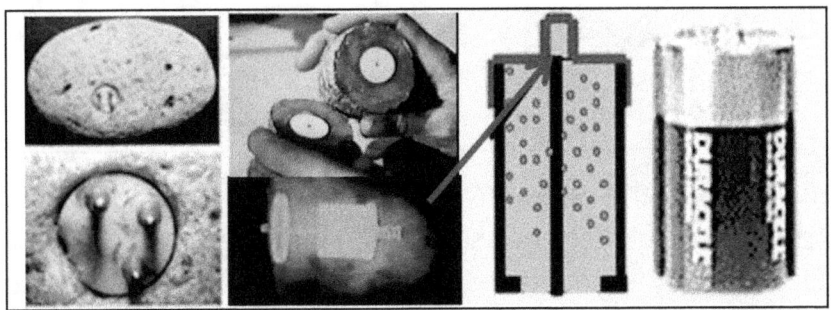

4000BC- Baghdad Battery- Then scientists found over a dozen ancient "copper-iron" batteries in Iran from about 6 thousand years ago. They would have provided about 1.2 Volts each for causing filaments to glow or electroplating [bottom row left].

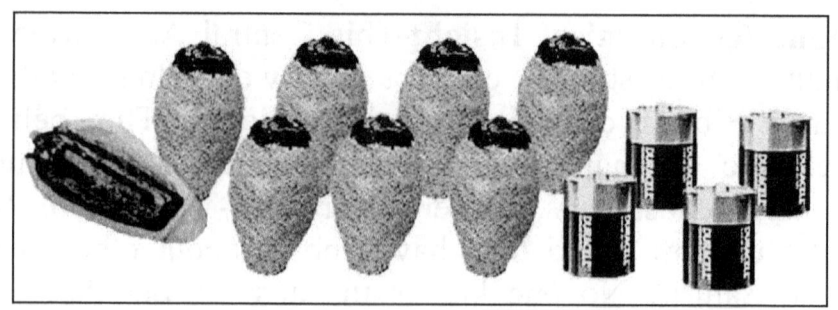

3000BC- Indian Battery-Then they found Sanskrit instructions in India for combining many "copper-zinc" batteries together to get 110VDC . The mixture was really zinc powder embedded in mercury making contact with a copper plate all in copper sulfate. With that combination the maximum voltage that can be obtained is only 1.5 Volts, so many "batteries" had to be connected as shown so that Indian electroplaters could do their work. The group below would only provide 13.5Volts.

Today we do the same thing. Hopefully you can tell that electricity was in use thousands of years ago and continued to be used. In my fantastic wisdom I made this assumption. If batteries were used around the world, someone may have known about electricity.

Anak in West Virginia-The picture to the right [following] has been determined to be a floor section that is thousands of years old. The reason this is of interest is that it shows a remarkable capability to "Grow Stones". These stones were grown at such an ancient time, the central core is almost completely gone, but the outside lattice structure of this ancient floor can still be seen in the West Virginian countryside. These would have been floors and walls used by the ancient Anak rulers. The joints between each waffle pattern can be easily seen today. Each of the blocks is layered as though the blocks had been grown in place rather than being quarried at a distant sight. In the close-up shown, note how close together each of the blocks is positioned to adjacent ones. Not even a needle could pass between their interfaces. The interior of each of the floor stones was of a softer material and eventually was eaten away.

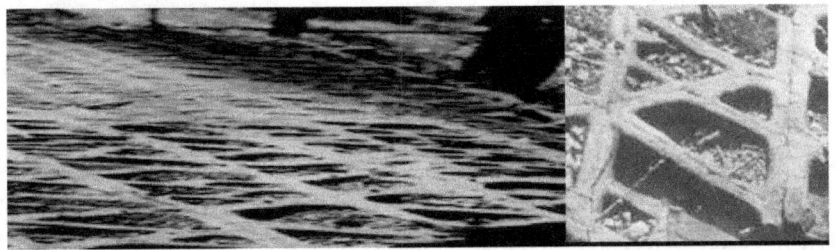

More West Virginia-The image below left, shows a West Virginian artifact that is very curious indeed. Note that the centers of the blocks are worn away and there is little or no space between the various "Rock rims" that are still visible. It is as if the rocks were covered in a hard substance that grew to meet adjacent rocks.

Anak in Oklahoma-We find the same thing in Oklahoma [Above right]. The "Grown blocks" in this Oklahoma Wall shown to the right, are believed to have been part of an ancient metalworking plant. If the blocks were not grown in place, where did the people find these perfectly matching stones that have not been externally shaped? We can tell the stones were not worked because the outer layer of ringed material on each block has not been violated. It is believed that some of these sites might have been processing plants for raw materials including Copper and possibly Uranium from the New Mexican fields. Additional sites show these same types of "grown blocks" [see below left]. Additional sites around the world including Peru and Australia assure us this was a common construction methodology of the Anak people.

  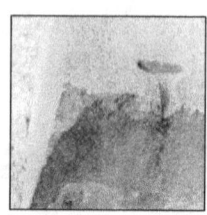 

Similar civilization in Brazil, Peru, Mexico, India, Egypt, and other sites show highly civilized pre-Cro-Magnon man communities around the world. Above middle is from a PreInca wall showing higher density rock along the outside of the stone and a very similar rock from Australia showing the layers from buildup.[Above right]

Manufactured goods- Investigators have found toys, jewelry, and construction tools from the Mesozoic Era. Some of the objects found include the bell, a workman's ladle, and even a hammer with a coalified handle.

Heavy Wheeled Transport-In Turkey was found dozens of tracks on Lava rock estimated to be from the Mid Tertiary period. The wheeled transports were so heavy, massive ruts in the rock are still extant. The tracks cut across the landscape of the Phrygia Valley as shown below.

  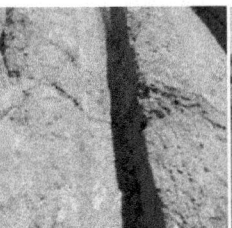 

Malta Ruts-Very similar interesting and mysterious tracks exist in other locations of the world, notably in the Maltese archipelago. These ancient grooves continue to puzzle researchers. Some of the strange tracks of Misrah Ghar il-Kbir deliberately plunge off cliffs or continue off land and into the ocean. It is not yet known who made the tracks shown below.

The islands of Malta and Gozo in the Maltese archipelago are scarred with hundreds, if not thousands, of parallel lines seemingly cut deep into the stone. These ancient grooves have puzzled experts for centuries. Some of the strange tracks deliberately plunge off cliffs or continue off land and into the ocean. Who made these enigmatic tracks, and why?

Huge Building Blocks-That brings us to many of the buildings from around the very ancient world. The Tower of Babel or Baalbek is a good example of what I'm talking about here. Built about 6 thousand years ago, the buildings simply could not have been built without heavy machinery or by giant people with capability to levitate to allow for reasonable block placements. The first image in the collage shows that unbelievably massive stones were so easy to place that the builders didn't put them on the bottom, second, or even third row. They placed them on the 4th and 5th level and these things are simply unbelievable at almost 1000 tons. A modern crane tried to move one and simply sank into the ground around the thing. By the way, each of the images below have tiny people in them. These are regular sized men and women.

Similar massive stone-works in Russia and Peru show these people lived everywhere. The giant race of men lived all through the Tertiary and Pleistocene age and into the Holocene, but many texts tell us of the final ending around 1000 BC.

Forensic Science-Some scientists started looking for holes and found bullet holes in skulls of Cro-Magnon and earlier people, pointing to the use of high speed projectile weapons used by the Anak people. Here are just a few examples.

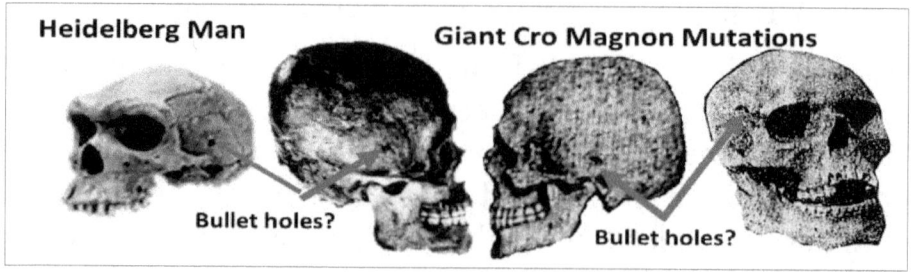

In Russia and extinct Auroch was found with a similar high speed projectile hole in its forehead which is believed to have been put there during the Pleistocene [See the following collage]. A similar hole was found in the same type animal in

Zambia and other signs of hostilities. Most had a tiny hole going in and much larger exit destruction. Please notice something that might be bad. Many of the bullet holes are from point blank range given the entrance hole positions as if done by a contract killer. I wonder if it was some form of mafia.

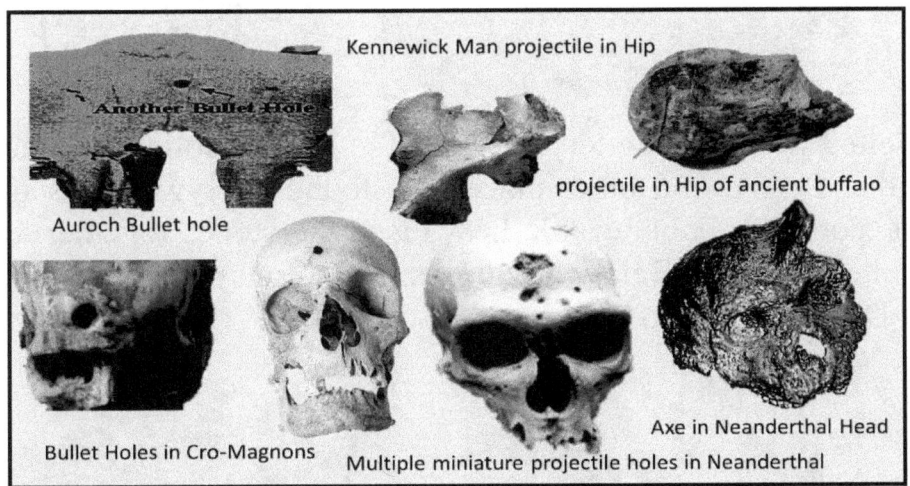

Nuclear War- Unfortunately, a sign of civilization is nuclear war and the nuclear evidence of the Bharata War 5500 years ago initiated by the Anak people. Radioactive skeletal remains of people who died suddenly in the ancient city called Mohen jo Daro, Pakistan show use of this unfortunate type of weaponry as hundreds of "huddled" humans are found in the now deserted streets show us Anak capabilities.

A massive area in the middle of the Libyan Desert has sand turned into miles of glass as if a massive nuclear explosion, not relating to a meteor, somehow was initiated 6 thousand years ago. The map shows the massive explosion point and one of the millions of pieces of melted desert sand produced.

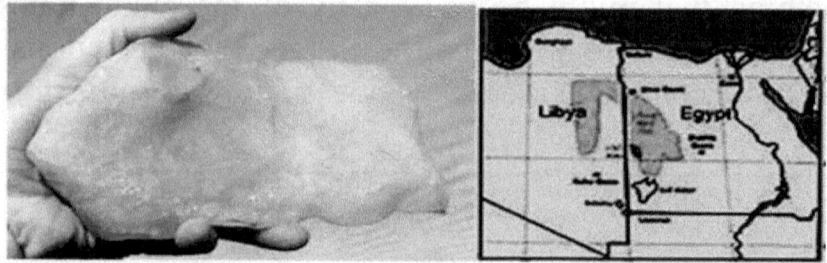

Melted walls and pottery in Mohen jo Daro, Scotland, France, India, U.S.A, and other places are found everywhere as if a weapon with unbelievable heat fused the stones together and melted the clay. Below are the remains of clay pots in Mohen jo Daro 6 thousand years ago

Enough processed nuclear material was determined to be missing from the 16 ancient nuclear plants located in the Oklo Mountains of Gabon, Africa to power New York for a year. Mysteriously, tons of the Plutonium are missing and have been missing for thousands of years. The image shows location of the various processing area.

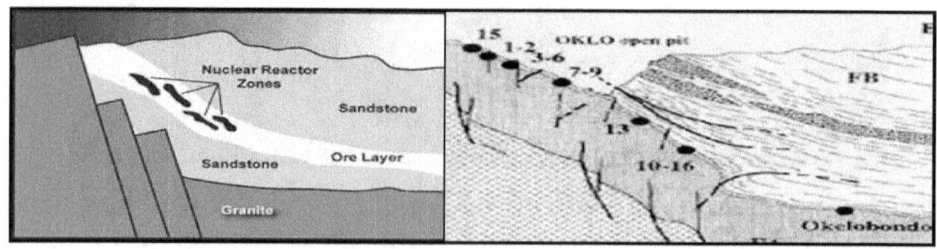

Three gargantuan walls were placed to protect a strange triple circular building in Bolivia. The building looks very much like a modern nuclear generating plant is still extant. The images below are of the 16 foot triple wall protection area in Sacsayhauman, Bolivia around the mysterious triple walled circular enclosure is shown below left and a similar nuclear processing plant in Russia is shown to the right.

Oriental Flight- Another thing that describes a high level of civilization is commercial and military flight. Bronze airplanes models have been discovered in China, too. Bronze models were found in ruins in Liaoning from 2500 years ago. While some try to say these are fish images, the triple wing tail gives them away. They are clearly jet planes very similar to models found in South America. These objects seem to be beyond the imagination of people in olden times. We can presume that they were not beyond their imagination because they saw them in the air and wrote about flying machines..

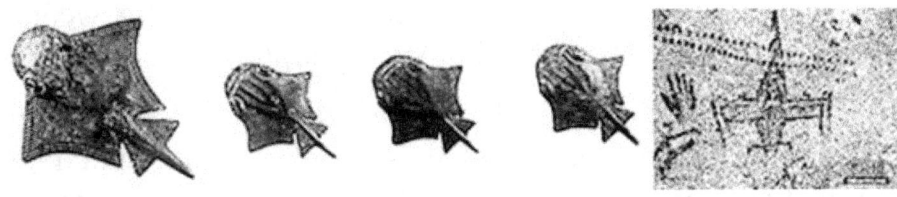

Another image was found on a cave wall [above right] and three more planes models were found similar to the jet models. [Next left]

In Chinese caves we also found similar flying saucer depictions to those being seen today. [See previous right].

2000BC, Chinese built some- Emperor Tang wrote about a *"flying chariot that was built and flown. The craft was destroyed to keep the secrets away from outsiders."*

French Flying-Below left is an image found in Niaux near the Pyrenees, France from well over 3 thousand years ago. In the middle is another image showing a sky battle and a pilot wearing a space suit of some kind from Peche Merle Cave, France.

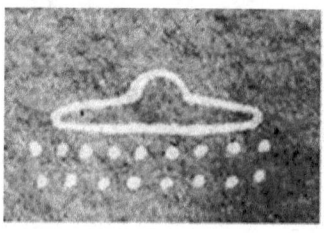  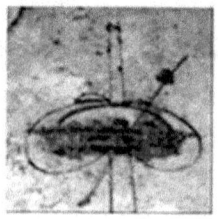

Italian Flying-The last image was found in a cave near Pompeii, Italy that was estimated to be 5000 years old. Notice that the ship has a 3 person crew.

*Greek Legend-*In discussions about battles between the gods we find the following: *"Hot vapor lapped the titans, flames unspeakable rose bright to the upper air [outer space], lightning blinded their eyes."* Apparently lightning weapons were used in outer space by the Anak people.

Pilots from Nepal-In Nepal, an indication of a strange visitors was found. A carved plate has been found which was manufactured around 4 thousand years ago. The carvings on the plate shows a large headed being similar to that depicted as one of the UFO pilots currently seen an elliptical shaped object above him, which is very similar to reported UFOs of today. That is odd enough, but what I want you to see here is that it looks like the giant man is traveling from Mars down to Earth in this football shaped thing. There also is a monkey looking thing coming from where Venus would be and a Lizard thing possibly leaving Mars, but let's mostly look at the Martian.

 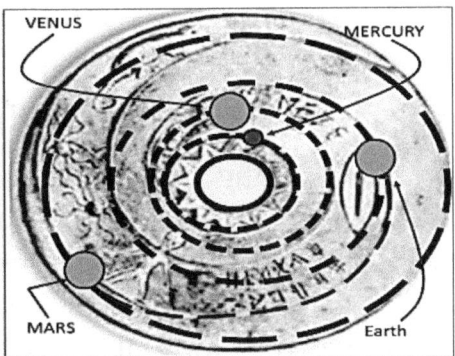

Iraq Flying- A ceramic model of what appears to be a three engine space shuttle, almost exactly like our new one was found in a Babylonian exhibit from 3000 years ago. The space shuttle has the same type 3-rocket ports out the back as our newer one, but notice the pilot's head has been lost. I put some fire out the back to show it in action. Yes, I also added his head .[See Next]

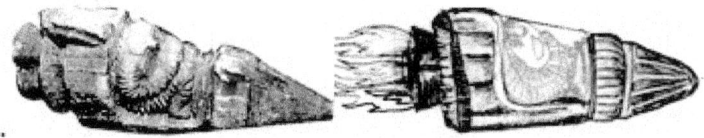

Japanese Flying Machine-In Japan we found a cave painting that shows a rocket ship which was estimated to have been drown well over 2000 years ago. [Below left]

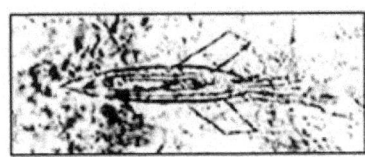 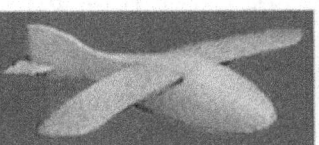 

Egyptian Description-1500 BC- As recorded, The Pharaoh Thutmose III saw *"silent, foul-smelling circles of fire and flying discs in the sky"*. Several models of Anak Egyptian airplanes were found in 1898 with a vertical rudder like modern aircraft were placed in the Cairo Museum and estimated to be from about the same time. [previous right] Besides having the ability of design and build flying machines, we are told these early scientists modified animals.

Genetic Manipulation

Flying, Growing blocks, levitation, nuclear weapons, highly skilled manufacturers, and all the rest clearly show a level of civilization, but what concerns us the most here is the art of turning an ape into an ape man or modifying humans. In this section we will look at one of the scientific achievements of the Anak called Genetic manipulation. I just read the other day that they are allowing human cloning as our geneticists go into some dangerous waters. The Anak were making all types of animals and modifying apes and ape-men and locating them all over Africa, and much of Asia until the great Homo-Erectus appeared and they began the same thing on them tying to make them a useful servant. We don't have to look far. Let me just give you a few to reinforce this important part of tracking the Homo-Erectus. If you ever wondered how some popped up in the most unusual places, the answers have been in ancient texts all along. We can imagine that Homo-Habilis and the Homo-Erectus initially had hair all over their bodies. But around the world the same description is found over and over again.

Indian History- *The first humans were covered with thick hair. When they mated [with Anak] they produced people as they are now.* [This doesn't mean there was sex involved, but we can imagine gene splicing was a popular science of the day.]

Sumerian Gilgamesh Story--*Aruru [God] pinched off some clay and created a [primitive man] Enkidu. His whole body was covered in hair. He knew neither people nor country, with cattle he quenched his thirst, a hunter and brigand—She [Shamhat —one of the Anak or Annunaki as the Sumerians described them] must take off her clothes and reveal her attractions. Do for the primitive man, as women [Annunaki women] do. She pulled not away, Enkidu was aroused.* [This is the most descriptive and shows that there were female Anak that had sex with male humans or artificially integrated DNA chromosomes, but the outcome was the same..] --*Afterward- the gazelles saw Enkidu and scattered, for Enkidu had stripped---* his body was too clean [the hair was all gone]. *His legs were diminished-he could not run as before, he had become wiser.—Enkidu, you have become like a god [Anak]- He shall bring up daughters of gods* [hybrid men]. [The Union between the Anak and the new human produced viable offspring.]

Mandaeans of Iran Story-*According to their traditions, the gods first made man. When he was finished, he looked like a man, but moved on all fours, had the face of an ape, and made noises like a sheep. Only later did he put in a soul and teach him and make him erect.* [This ape-man was not Adam.]

African Story- *In Africa, the same story was told. Hairy men became human after coming in contact with Anak.*

Southeast Asian Story-*An extremely hairy human "female" named Bota Ili was cooking food. A non-hairy fisherman named Wata Rian saw her and got her drunk. While she was asleep, he shaved her entire body. Only then did he find out she was a woman. She learned to wear clothes, they married, and they began a new race.*

Emerald Tablets [Egyptian Story]-*The master said-take them by the arts ye have learned of far across the waters until ye* reach the land of the hairy barbarians, dwelling in caves of the desert. Follow there the plan. [The plan was to inbreed with the hairy barbarians.]

Ngombe Tribe-*See if the Ngombe tribe folklore doesn't sound familiar. "A sky person [Anak] saw a hairy man [6th era man]. She married him and removed his hair. Then a Garden was made for man to live in."* [The offspring of this union was less animal like just like all the other descriptions.]

Inca legends-*The age of primitive man [hairy man?] was before the age of heroes [Anak and demigods]* - [Primitive man turned into heroes probably by a process called genetic manipulation.]

Aztec and Mayan history-*According to Codex "Laticano-Vatino" the first man [Homo Erectus man] was hairy. During the age of the four winds men turned into monkeys.* [This may be a reference to man with thick hair or something else we will look at later.]

Mongolian Verbal History-<u>*God created a man*</u> *and woman out of clay. Their entire bodies were* <u>*covered with a layer of fur.*</u> *[Sounds like homo-Habilis] This was during a time when the seas were still rising. He went to get some "everlasting life water" and ordered a dog and cat to watch over his new creations. The devil gave the animals some milk to distract them while he urinated all over the humans. God was angry with the dog and cat for not caring for the humans. As punishment, God made the dog lick off all the hair from the human body except for the area around the groin and under the arms. After each lick, God placed the fur taken away and placed it on the identical area of the disobedient dog.* [OK this one is weird, but it does indicate Homo-Erectus was now without much fur.]

Animal Abominations

Around the world and in our Bible we read about unclean or "abominable animals". The reason most of the animals of this time were considered abominations was a massive thrust for geneticists to modify animals. Besides our Bible, we find out from other Jewish texts the same thing as sceintific research in Genetics and Engineering bring us to a dangerous level. I know this sounds like some bizzare fiction, but bear with me for a minute as we review a tiny portion of the Judeo-Christian texts concerning this "seemingly" erroneous fact described in the Bible about what the Anak were cabale of producing.

Book of Giants [Ancient Essene Jew text]- *- they knew the secrets of heaven and sin was great in the Earth. They made mistakes and they killed many [animals and people- They selected two hundred] donkeys, two hundred asses, two hundred rams of the flock, two hundred goats, two hundred other beasts of the field. From every animal, and from every [type of human was taken its seed] for mixed sex. [After a time] they defiled the animals **and people** and begot giants, monsters, and dragons.*

Book of Secrets [Ancient Essene Jew text]- *Those who would penetrate the origins of knowledge, along with those who hold fast to the wonderful mysteries of life. -With this I beseech your attention. All of the secrets of manipulating life were known but they [the ancient humans] did not know the secret of the way things are nor did they understand the things of old. Belial who modified creation, a thing that ought never to be done again, except by the command of his Maker. You have not become wise in understanding my secrets of life and the earth; for you have not properly understood the origin of Wisdom.*

Jasher 4:16-18-[Canonized Jewish text] *and the sons of men in those days took from the cattle of the Earth, the beasts of the field and the fowls of the air, and taught the mixture of animals of one species with the other, in order therewith to* provoke the Lord; and God saw the whole Earth and it was corrupt, *for all flesh had corrupted* its ways upon Earth, all men *and all animals*. ---And after this *they sinned against the beasts and birds, and all that moves and walks on the earth:*

Enoch 2:4-5 -*[Canonized Jewish text]* And the sons of men went and they served other gods, -- and the sons of men forsook the Lord all the days of Enosh and his children; and the anger of the Lord was kindled on account of their works and *abominations which they did in the earth—*

Enoch 4:8-And *lawlessness increased on the Earth [War]* and all flesh corrupted its way, alike men and *cattle and beasts and birds and everything that walks the Earth all corrupted their ways and their orders.* The only way that animals corrupted their way was that they were genetically manipulated and just weren't the same animals.--

Enoch 7:5- And they began to *sin with birds and with animals and with reptiles, and with fish*. [This did not mean that the rulers had sex with fish. This is talking about manipulation of species]--

Jubilees 5:3--*[Canonized Jewish text]* -and all flesh corrupted its way, alike men and cattle and beasts and birds and everything that walks the Earth all *corrupted their ways and their orders.* [The only way that animals corrupted their way was that they were genetically manipulated and just weren't the same animals.]

Jubilees 7:24- Afterwards *they sinned against beasts and birds and everything that moves or walks* upon the Earth. [There are two ways to sin against beasts- sex and genetic manipulation. God didn't like either.]

Enoch II 59:5-6- But whosoever *kills a beast without wounds, kills his own soul and defiles his flesh*. And he who does any beast any injury whatsoever, in secret, it is evil practice, and he defiles his own soul. [The killing and injury done in secret was not killing animals for food, it was the genetic manipulation and corruption by integrating man's; genetic material.]

Generations of Adam 6:1-10--[Ancient Gnostic Jewish text] AMONG our little ones - <u>Timnor's sister Ammah</u> was also blessed with understanding, for she investigated the nature of life, unlocking the mysteries of life itself. -- Ammah was not one whit behind her husband in creating wickedness, for <u>she manipulated the very fountain of life, until she had created new forms of beings dedicated to evil</u> and the destruction of mankind. Later we find that Timnor created great machines of war including flying machines and vehicles that went under the Ocean.

Generations of Adam 8:4-20 Timnor and Ammah practiced every abomination. Tranter learned the ways of his mother Ammah and he did manipulate the very nature of man and beast <u>to create new forms which God had not ordained.</u>

The Zoroastrian "ZAND-AKASIH"[Bible] - <u>Satan miscreated creatures and they became useless.</u> <u>God saw the defiled and bad creatures, they did not delight Him</u> [They became abominable]. Satan's downfall was the <u>unrighteous creation of the creatures</u> and ignorance.

There are many other texts describing how the Anak with the assistance of the Neanderthal and the Cro-Magnon to follow had just gone crazy making new "unclean" animals including new dinosaurs. They also used them for target practice.

Dinosaur Fun

Today we have fantasy movies about Jurassic park/ remade animals and we are finding out that during the Pleistocene Age, many animals were not only made by gene splicing, but also ancient animals were re-made to be abominations to God. We are finding dozens of unfossilized dinosaur remains showing they were re-made less than 40 thousand years ago even when it has been determined that the Duckbill dinosaurs, Tyrannosaurs Rex, and other unfossilized dinosaurs did, in fact, become extinct by the end of the Cretaceous. Someone must have found viable DNA almost like the Jurassic Park movies. Speaking of wrong animals in the wrong era, check out the next section. Now the world is being flooded with soft tissue, dinosaur blood, dinosaur skin, and all sorts of unfossilized dino-stuff. Some of the blood and tissue samples are next.

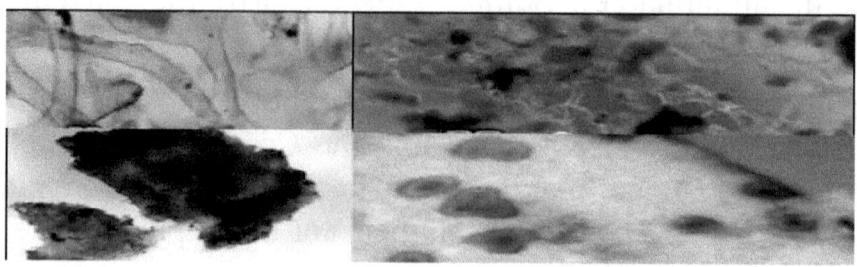

Besides blood, researchers are finding all sorts of pliable, soft tissues. Some colleges have even tried to recreate the same type of experiment identified in Jurassic Park movies. Following are more samples.

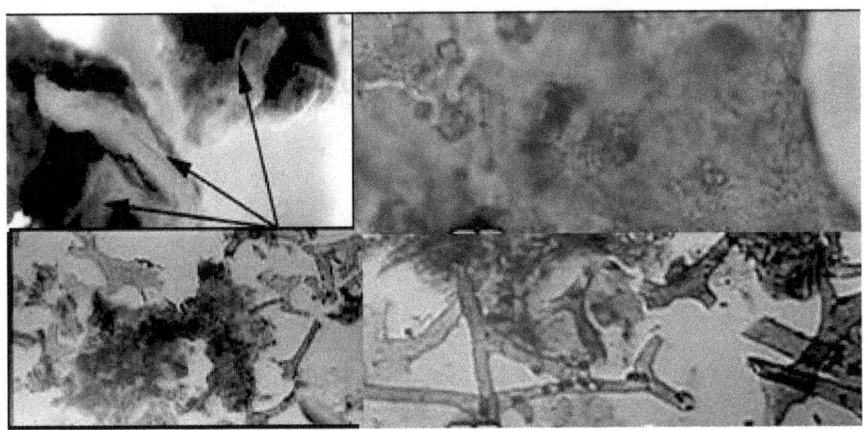

Apes Before Dinosaurs?-Why does the ape skeletal remains seem to have such a horrible look on its face? [See next left] The reason is obvious to a casual observer, but hard to accept by the ardent evolutionist, because, the fear experienced by this mammal was the fear of being eaten by an allosaurus well before it could have become evolved to the high form of the ape or it was a remanufactured dinosaur. The find was in upper New Mexico, but typically, you don't see these things in text books. One of the photographs of this impossible scene is shown below and no, it was not produced with trick photography.

Entropy of Evolution-Just think about this. If scientists would just read some of the ancient texts they would not have to hide the anomalies any longer and there would be a logical reason why people remembered seeing dinosaurs and painted then, carved them, and etched them into wall.

The images above from Palestine, Greece, Utah, and Colorado, Guatemala, Peru, Mexico, and Cambodia show manmade dinosaurs were one the earth even past the Pleistocene Extinction. An additional group from Peru, England, Spain, Babylon, Israel, and Egypt show that dinosaurs were not rare for a while.

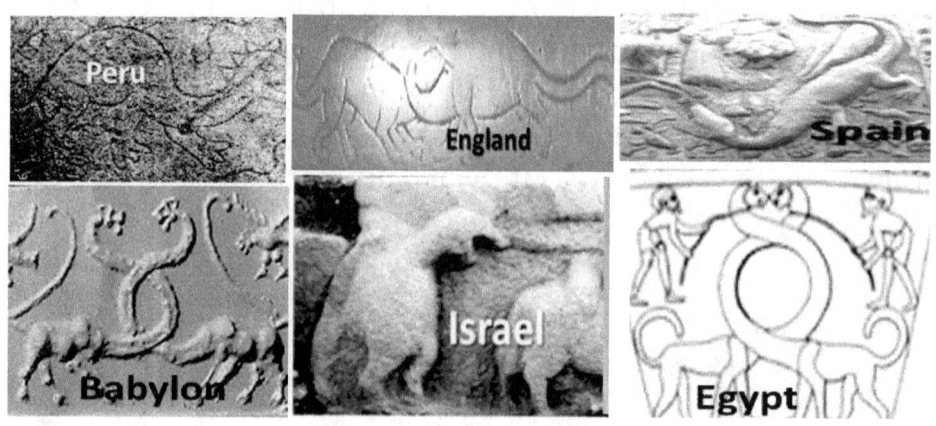

Then there is the book of Daniel in the Bible. This is the last chapter in the King James Version of the Bible. In it King Cyrus thinks his captured dinosaur is a god that can't be killed so Daniel kills the beast. Later versions have taken this part out of Daniel because they simply didn't know about all the other evidence.

Daniel [Bel and the Dragon] 1:23-28- *And in that same place <u>there was a great dragon, which they of Babylon worshipped</u>. And the king said unto Daniel, Wilt thou also say that this is of* *brass? lo, he liveth, he eateth and drinketh; thou canst not say that he is no living god: therefore worship him.-Then said Daniel unto the king, I will worship the Lord my God: for he is the living God. But give me leave, O king, and I shall slay this dragon without sword or staff. The king said, I give thee leave. Then Daniel took pitch, and fat, and hair, and did seethe them together, and made lumps thereof: this he put in the dragon's mouth, and so the dragon burst in sunder: and Daniel said, Lo, these are the gods ye worship. When they of Babylon heard that, they took great indignation, and conspired against the king, saying, the king is become a Jew, and he hath destroyed Bel, <u>he hath slain the dragon</u>, and put the priests to death.*

If you get anything out of this book, I hope it is that genetic manipulation was commonplace during the Pleistocene Age and many successful animal types were created. This would have included the manipulation of Apes to form ape men and the manipulation of humans to "Modify" them. Maybe a chart of two will help.

Curious Charts

I know some are thinking this is way too strange, so I though a couple of charts will help you. First off, there is no question that evolution occurs but in order for modifications to not push a species further and further into disorder [Law of Entropy], some outsider MUST modify the chain of events to assure advancement of a species rather than the slow destruction of entropy. From the text I just presented there is little doubt the Anak people had violated the laws of God by "miscreating". Some of the miscreated were considered human. Most of the miscreating was just biologists having fun; completely unconcerned about the outcome. If you look in Encyclopedia Britannica, they present the first chart shown next as the development of animals and when they occurred. Essentially they stated that for millions of years there was about 500 different families of animals.

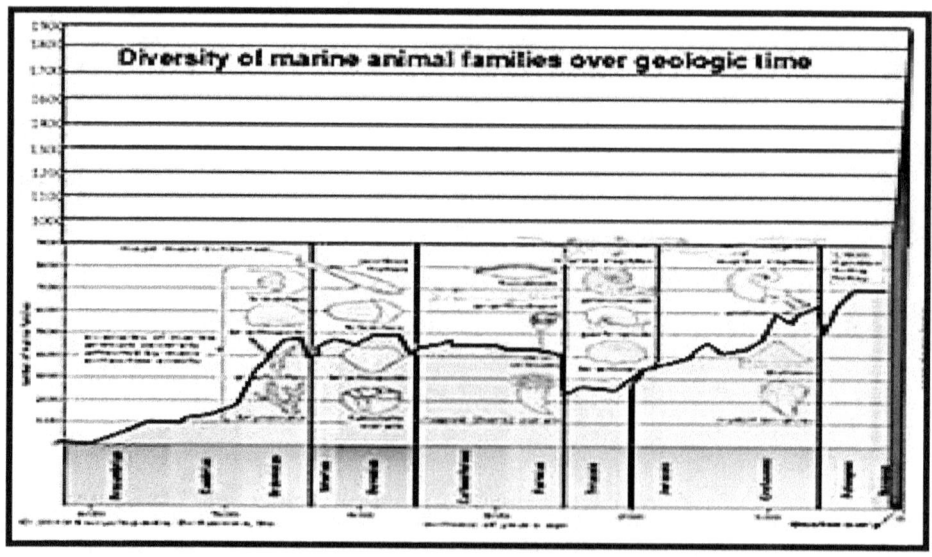

While there was a slight increase about 40 thousand years ago they tell us BAM 1600 families of animals miraculously appeared in the blink of an eye. Oh yes! We need to look at the new timing method because nuclear decay was so wrong, so I modified the chart using the new timing. Guess what? We now have 500 families of animals over hundreds of thousands of years and then BAM! 1600 almost overnight as if some massive group of genetic engineers were engineering away just like we are starting to do today. There can be little question about the ANAK miscreating all types of animals including the remanufacture of dinosaurs and the manipulation of Apes and humans.

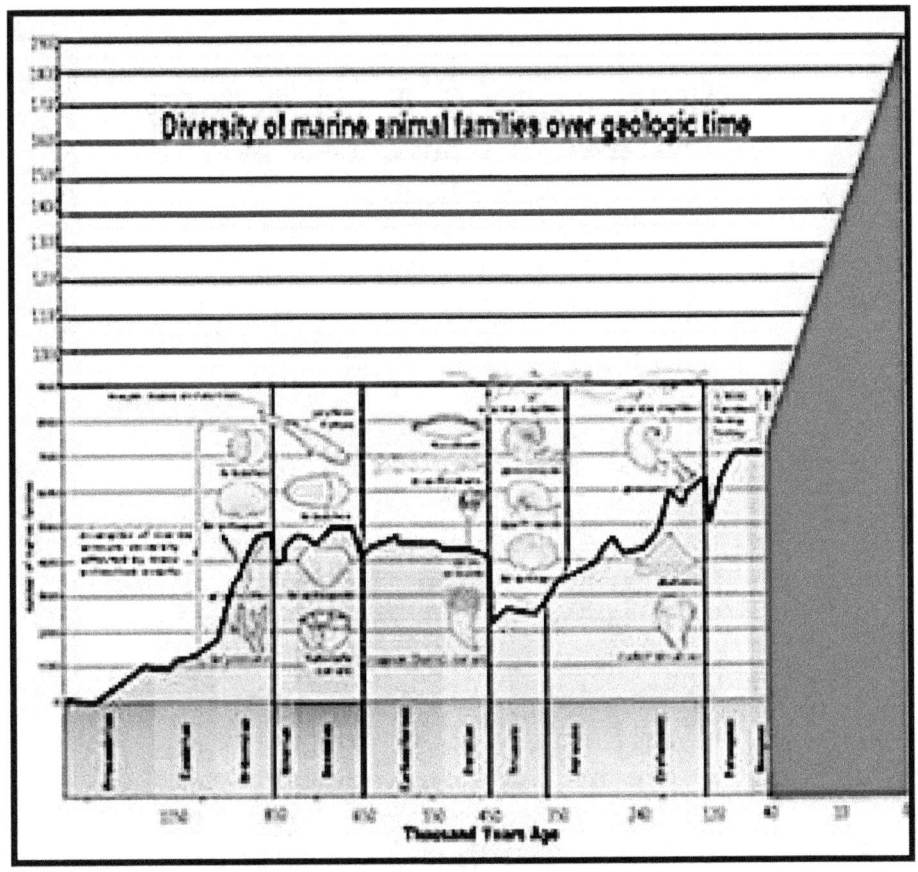

Human Mutation Chart

While we are on strange charts let me briefly look at DNA mutations. The chart following is the general details of major mutations of the Y-Chromosomes of humans. DNA Haplotype scientists record the combination of these mutations and when each one occurred by looking at the DNA. In this case, the "Y-Chromosome" track of DNA in the nucleus of each cell tells them your family history back to the first "human". For most people this track goes back 40 thousand years, but for individuals with a strong black background, this sequencing goes back an estimated 100 thousand years. While the first 4 mutations are during ancient times, 3 of the 4 are only shown in Black people. All that is to say this. Let's look at a track of mutation timing.

Homo-Erectus
A="Y-DNA Adam" [100 thousand years ago]
B= Sapien [50 thousand years ago]
Cro-Magnon
F= Adamic [40 thousand years ago]
C= Negriod [20 thousand years ago]
D= West African [12 thousand years ago]
E= Nubian [12 thousand years ago]
G= Armenian [12 thousand years ago]
K= Japheth/Asian [12 thousand years ago]
H= Afghan [12 thousand years ago]
I= PreGreek [10 thousand years ago]
J= Hamite [10 thousand years ago]
T= Near East [6 thousand years ago]
R= Scythian [6 thousand years ago]
M= Russian [6 thousand years ago]
O= Oriental [6 thousand years ago]
P= East Europe [6 thousand years ago]
Q= American [5 thousand years ago]
S= SE Asia [5 thousand years ago]
L= Dravidian [5 thousand years ago]
N= Scandinavian [5 thousand years ago]

The following chart shows all major mutations of humans since humans were first established 100 thousand years ago. Do you notice anything odd?

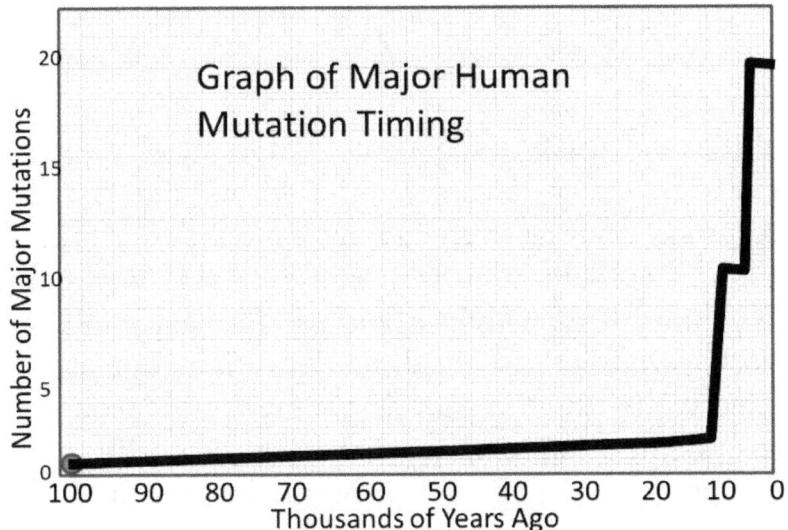

This is not the chart of random mutational events. This shows that humans did not mutate at all for almost 100 thousand years, then something happened, and then something happened again to change humans. The changes can most easily be understood if we assume they were helped by nuclear events and/or Anak genetics.

One reason it is hard for some to accept the thousands of bits of evidence concerning th eAnak people and their influence on the development of huamns is that there aren't any more living so their existance can easily be swept under the rug.

End of the Anak

We find that the Anak descendents would survive until about 1200BC when the Adena generally killed the last ones in North America, and the PreInca may have killed off the last of the Akamim in South America. While all this was going on, The Jews under King David killed off the last of these giant people in the Middle East. And the Fomorian giants of Ireland were killed off by the "Tuaath de Dannan" descendents of the Scythians. While Irish history is somewhat colorful, here is an excerpt of how the last of the Fomorian giants died.

King Nuada was the leader of the Dannan, but he lost his arm in battle so he found a doctor that replaced it with a working silver one. With his new more powerful arm, he led the Dannan into battle with the Fomorian giants. Nuada was killed by the Fomorian King Balor's poisonous eye, so Lugh, of the Dannan, killed Balor and took over as King of the Dannan after destroying the giants.

I know David killed the 11 foot tall Goliath with a sling and a stone, but this is more hero-like. To be honest, the reason all of the heroes did so well defeating the Anak around the world is that they were of extreme age and feeble by this time. Let's take for instance when Moses Killed his second Anak giant who was King of Bashan. His name was Og and he was also about 11 feet tall. This excerpt comes from the Ancient work "*Book of King Og*".

Book of Og- [This is Og talking to Moses just as Moses entered his castle to kill him..]- *I, Og and the true Rephaim [one of the Anak clans] - tore out the jelly of their eyes and* *pulled the stupid skulls off of those false Rephaim priests of in*complete loin [He hated circumcision.]. *We spilled their worm-filled stomachs and* yellowing *wet* intestinal *droppings out on* the fields. *With bare hands we twisted and* tore them free. *The proper use of smaller selves* [regular sized people] *were as slaves, temple servants and food -The smaller self men indeed became food for the Giants and their male offspring were indeed burnt in sacrifice. I have watched your mother* Egypt - *eat your young. I watch now and I ponder is this why you now war with me, O worm of* Israel? *You have slain my neighbor in Sihon.* [Moses had already killed one of the Anak] *I Og will tear the virgin child skull of the corrupt* fecal worm *Moses from his soft dung-filled body-I want no victory spoils. Just your stupid head.-I have trained for battle with* strange *weapons.-Make war little man. Make war corrupt little worm. For Baal of the earth I'll spread the blood of your virgin body.-- I Og have been training each arm as one trains their oxen. My* old-world *back and legs are ready. My life is to destroy you* fecal worm. *I will tear your backbone out* your backside. *For Baal of the earth I Og will pull the* fecal-bearded *smaller self child skull of Moses from your* dung-filled *body. I Og the* last of the Rephaim *know how long Moses has to live.*

These were his last words according to this ancient work as Moses killed this Anak human who was thousands of years old by this time. Possibly one of the reasons for his being killed was to shut up his incessant cursing. No doubt their feebleness in old age played a role in their final ending. While they were in control, possibly the first experiments were to modify the Australopithecine apes into something different.

Habilis-like Ape-men

Possibly the first experiment produced what is now called Saheanthropus which was followed by Paranthropus, Rudalfensis, and finally Homo-Habilis. None of these were true humans in the strict sense.

Sahelanthropus Ape-man

Sahelanthropus is one of the oldest known species in the human family tree. This species lived about 6 million years ago in Central Africa by the old dating system or about 100 thousand years ago by the 4 new dating methods I mentioned. Walking upright may have helped this species survive in diverse habitats, including forests and grasslands. Although we have only cranial material from Sahelanthropus, studies so far show this species had a combination of ape-like and human-like features. Ape-like features included a small brain (slightly smaller than a chimpanzee's), sloping face, very prominent brow-ridges, and elongated skull. Human-like features included small canine teeth, a short middle part of the face, and a spinal cord opening underneath the skull instead of towards the back as seen in non-bipedal apes.

 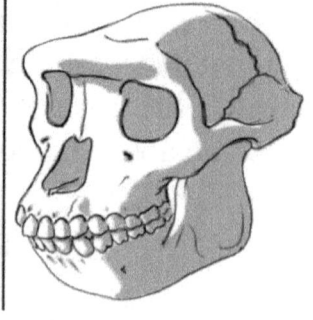 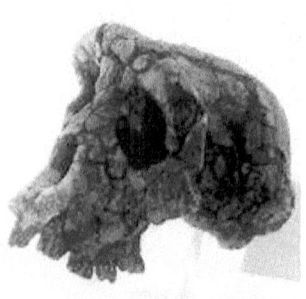

Paranthropus Ape-men

Paranthropus were bipedal hominids that probably made from the australopithecine hominids 2.7 million years ago in central Africa by the old dating or about 95 thousand years ago by new interpretation. They are characterized by robust craniodental anatomy, including gorilla-like sagittal cranial crests, which suggest strong muscles of mastication, and broad, grinding herbivorous teeth. However, Paranthropus skulls lack the transverse cranial crests that are also present in modern gorillas.

Homo-Erectus Naledi

While this guy is called Homo-Erectus, he was not. In 2013, fossil skeletons were found in South Africa. As of September 2015, fossils of at least fifteen individuals, amounting to over 1500 specimens, had been excavated and tentatively determined to be from a group of Homo-Habilis like hominids living from 2.5 million years to 900,000 years ago in the old dating method [90 to 80 thousand years ago-new timing].

 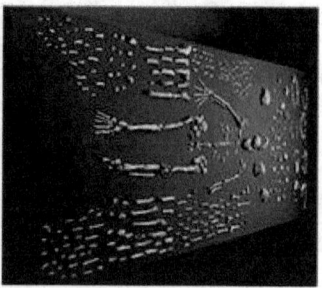

The strange thing is that a nearly complete hand and foot was found that had amore humanlike features while the arm looked more Australopithecus-similar shoulder and fingers and a Homo-similar wrist and palm. Even more strange, the structure of the upper body seems to have been more apelike. Four skulls have been discovered with cranial volumes of between 460 and 560cm^3 which places it as Habilis-like.

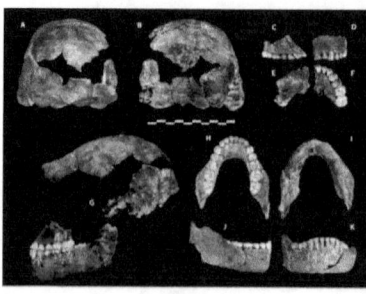 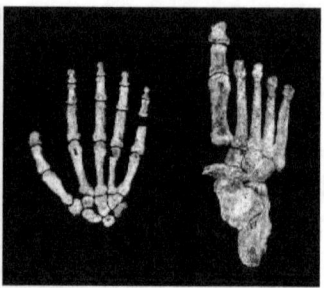

These guys were taller than a normal Habilis. Adult males are estimated to have stood around 5 feet tall and weighed around 100 lb. The hip appears to show naladi stood upright and was bipedal as its legs, feet and ankles are more similar to Habilis. The artist rendition shows the difference in Australopithicus, Erectus and this in between ape-man.

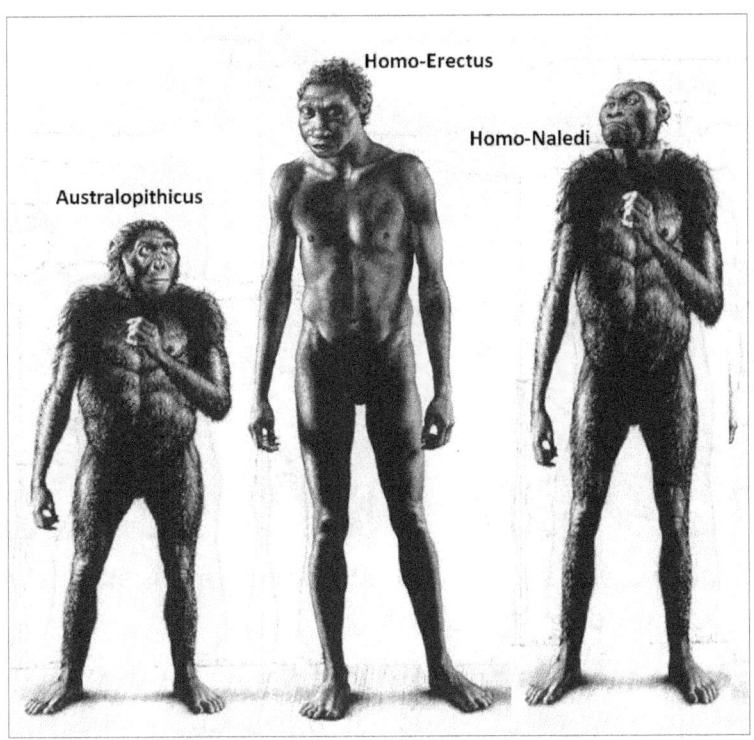

Homo-Habilis Ape-man

Homo Habilis lived during the Pleistocene approximately 2 to 1 million years ago by the old dating system [about 90 thousand years ago by modern techniques. While he was called HOMO- most place him Australopithecine like the other two experiments. I'm just putting him here for completeness. He was the most advanced ape-man of that time.

Body: Habilis was short and had disproportionately long arms compared to modern humans.

Neck: He didn't have the powerful neck muscles, or the large occipital opening at the base of the skull, so he was much more comfortable walking monkey-like.

Thigh: His thighbones were curved like other apes.

Face: He had a less protruding face than australopithecines but had a cranial capacity slightly less than half of the size of modern humans with a brain size of $600 cm^3$.

  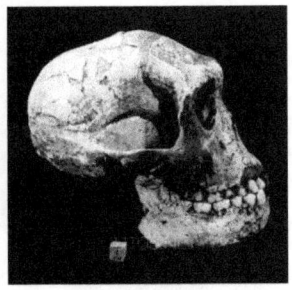

Appendages- Hands: The Habilis didn't have an opposing thumb Feet: His feet were still hand-like. While having hands on your feet might be interesting, it is not a sign of the advanced "Humans".

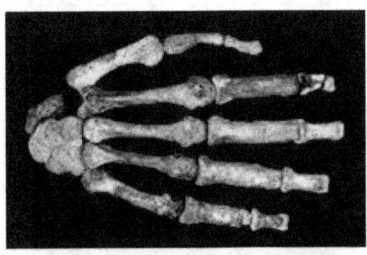 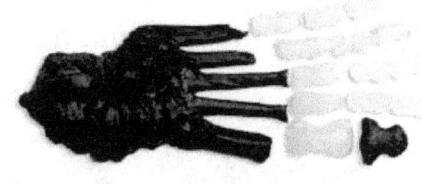

Tools: He made and used primitive stone tools.

Rudalfensis Ape-man

Like Homo-Habilis this was not a human and he also appears to have been a modification from Australopithicus. Skeletal remains from northern Kenya had been found: two jawbones with teeth and a face. The face was of a juvenile, but had features in common with the jaws. Homo Rudalfensis was born. Many believe this guy was also not human just like Homo Habilis.

Like many of the others in this grouping, Rudalfensis lived about 1.9 million years ago by nuclear decay or about 88 thousand years ago by newer methods. As Habilis and this guy lived about the same time, some have tried to determine a common ancestor, but to date these guys seems to just have sprung up around the same time.

 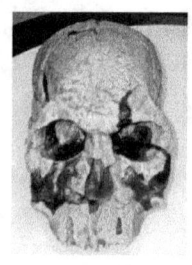 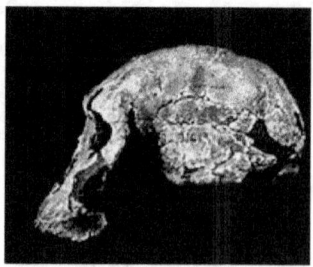

Skull: The construction looked very ape-like, possibly due to the large brow ridge. The face was said to be "incredibly flat", with a straight line from the eye socket to the incisor tooth. And the cranial capacity was about 700 cm^3 that was rounder than homo-Habilis and his teeth were larger while having a smaller overall head size.

Jaw: The jawbones were shorter and more rectangular than known Habilis. [See image above left]

Hands: The Rudalfensis didn't have an opposing thumb.

Neck: He didn't have the powerful neck muscles, or the large occipital opening at the base of the skull; so like the others he was not made for walking upright for any distance.

Feet: While we don't have much to go on, it is believed his feet were still hand-like and his thighbones were curved like other apes and the Habilis.

Habilis, Rudolfensis, and Paranthropus co-existed somewhere in the East African area between 2.0 and 1.5 million years ago [90 to 85 thousand years ago-new timing] while a new human came along who would be modified and found in many parts of the world. While the Habilis ape-men were almost all found in Africa, this new man was much more as he would not start in Africa, but instead first lived in Georgia, one of the Russian satellite states.

As the Anak had a new human to mess with, it seems the Homo Habilis group stopped changing and finally just disappeared. They were adapted to the environment, but the Anak lost interest. The next guy seems to be between Habilis and Erectus. While some placed this guy as Erectus, generally we can believe him tyo be a Habilis variant living in Russia.

Homo Georgicus

Out of Africa came a little faster than you were told in School Homo Georgicus was found in 2002 in Dmanisi, Georgia. This guy seems to be the earliest "true man" and appears to a variant of Homo Erectus. This shows that Homo-Erectus and Homo Ergaster were made from this first "seed".

A partial skeleton was discovered in 2001. The fossils are about 1.8 million years old by the old dating which places it about 85 thousand years ago like several of the others. Moreover, *georgicus* individuals are about the same size as the few adult *habilis* specimens known, and their skulls are rather similar. However, *H. habilis* is known only from sub-Saharan Africa.

 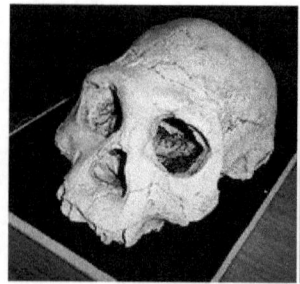 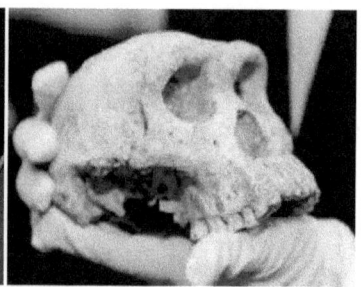

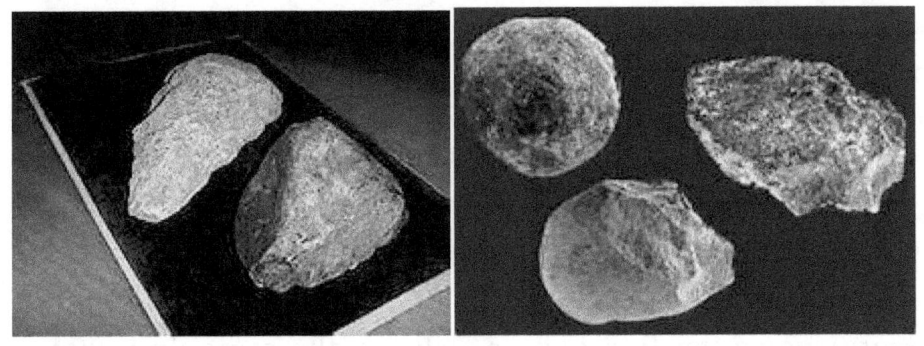

Tools: Some implements and animal bones were found alongside the ancient human remains.

Teeth: Tooth-wear patterns and remains found at the site show H. Georgicus had an omnivorous diet

Fire: There is no evidence of the use of fire.

Brain: At around 600 cm³ brain volume, the cranium was the smallest and most primitive human Hominid skull ever discovered outside of Africa until Floresienesis [hobbit] was found in 2003.

Dimorphism: A strong sexual dimorphism was noted. Males were significantly larger than females. This is considered a primitive trait and is much less obvious in Homo-antecessor, Homo-heidelbergensis and Homo-neanderthalensis type humans.

Size: Smaller than Homo Erectus [about 4 feet tall].

Outside Africa: Georgicus was the first species human to settle in the Middle East as the map shows.

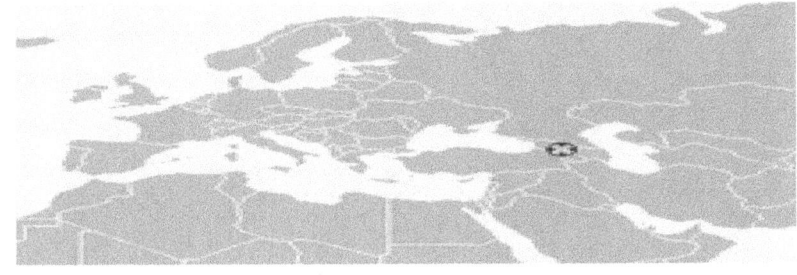

Skeleton: Four fossil skeletons were found, showing a primitive skull and upper body but with relatively advanced spines and lower limbs, providing greater mobility.

Classification: While Georgicus is sometimes considered a variant of Erectus, the skull showed shows marked differences. He had a smaller brain, but his skull ridge seemed to be much less pronounced. A number of researchers classify Georgicus as the first human. Almost the same time that this variant of Homo-Erectus was found in Georgia, we find the same human in Africa. Because of the tiny brain, we really should classify Georgicus as Habilis. Let's look at how Habilis compares to Erectus before getting into the variations of homo-Erectus and what that means to us.

Erectus Compared to Habilis

What we know is that there was a huge difference between the ape-man called Homo-Habilis and the human Homo-Erectus. The undirected Evolutionist quasi-Scientists can't understand what happened. It was not a massive electrical storm that enhced humaoids in opposition to Entropy. Here are some of the changes noted.

	Habilis-ape-man	Erectus-man
Location	East Africa only	Eurasia and Africa
Build	robust - 4 feet	Taller, slender -5 feet
Face	Protruding face, prominent cheekbones	flatter face, large brow-ridges
Limbs	long arms, short legs	Slender arms and legs
Walk	Feet outward	Feet more straight
feet	Hand-like	More foot-like
Posture	Stooped	upright posture
Teeth	Large and elongated	Shovel-shaped, smaller
Brain	600 cm3	1000 cm3
Tools	Scavenging tools	Hunt & defense tools
Fire	No use	fire and fire-making
Speech	no	yes

Homo-Erectus was the first true human of this line, but we find many variant of the Homo-Erectus before getting to the Cro Magnon. His features were man-like including his teeth, pelvis, and legs. He was much larger and his brain had swelled to <u>almost twice</u> that of his very recent predecessor. Those things didn't happen by chance and they all point to a very strong, very articulate, very manlike worker. The brain expansion was almost like the brain had gone into an evolution jet going thousands of years in the future while the same general shape and appearance looked almost like a reasonable progression. Don't confuse evolution with NEW CONSTRUCTION. I know you are going to say what is scientific about that? Well the undirected evolution idea that many types of sugars just happened to be on a stump at the same time and lighning struck forcing the sugars into clumps of DNA while on another dozen stumps the exact same thing happens in exactly the same way so that a viable procreative entitity could be established sounds much more fanciful. Speaking of fancy, let\s go right into Homo-Ergaster.

Ergaster Erectus

While called Ergaster, this is the guy we usually think of as Erectus. Homo Ergaster is believed to have been the first human by some. As such, he would have been the man described in the Genesis story that appeared during the 6th age after the end of the Cretaceous period and the "out of Africa Story" would start and end here. As shown below Homo Ergaster, was possibly the beginning of the Negroid Race. Certainly, it has come through many changes, but the Adam and Eve described as the 8th Age humans were NOT from this group.

Homo Ergaster simply means "working man" so I guess you can say there are a number of Homo Ergaster people living today. This guy is extinct and did not make it past the Pleistocene extinction. He lived in eastern and southern Africa during the early Pleistocene, between 1.8 million and 1.3 million years ago as depicted in nuclear decay. Modern timing has him living <u>about 100 thousand years ago</u>. Interestingly, Haplotype scientists tell us the first human Adam and Eve developed in Africa came about around 100 thousand years ago. This determination is by placement of the mutations along the DNA string rather than nuclear decay

There is still disagreement on the subject of the classification, ancestry, but it is now widely accepted to be the direct ancestor of later humans including the more well-known Homo Erectus although Ergaster and Erectus lived during the same time so there are many who do not separate them as different. I will identify both, but they might just be cousins without any real difference in mutated DNA. Without the DNA, we are only guessing. Some consider Ergaster to be simply the African variety of Homo Erectus who has been found at many locations outside Africa including Java and China.

Tools: Ergaster and Erectus both used more diverse and sophisticated stone tools than the Habilis. He refined the "Oldowan choppers" [next left] and developed the first "Acheulean bifacial axes" [next "2nd"]. OK; they weren't much, but sharpening stones took a knack.

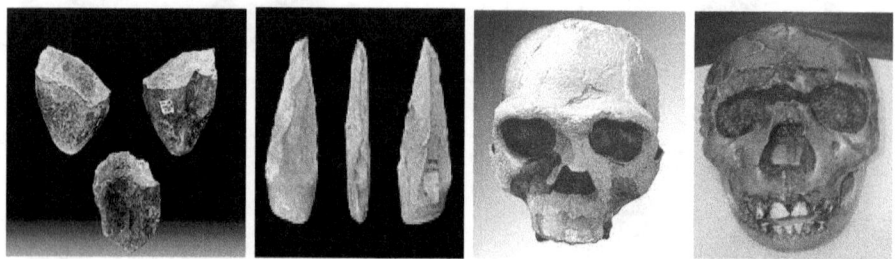

Fire: Ergaster was the first human to harness fire either by containment of natural fire, or as the lighting of artificial fire. This is still a matter of contention.

Linguistic: It was thought for a long time that Ergaster and Erectus were restricted in the physical ability to regulate breathing and produce complex sounds, but now they are believed to have been able to speak.

Anatomy: Let's say, Ergaster/Erectus is the oldest known early humans from which modern man would descend. He had relatively elongated legs and shorter arms compared to the size of the torso. These features are considered adaptations to a life lived on the ground, indicating the loss of earlier tree-climbing adaptations, with the ability to walk and possibly run long distances.

Brain: Compared with earlier fossil humans, note the expanded braincase relative to the size of the face.

Teeth and Growth Rate: The most complete fossil individual of this species is known as the 'Turkana Boy', dated around 90 thousand years ago using newer dating methods. Microscopic study of the teeth indicates that he grew up at a growth rate similar to that of a great ape.

Society: There is fossil evidence that this species cared for old and weak individuals.

Area of Civilization: This guy has only been found in Africa.

Size: Ranges from 5 to 6 feet

Weight: Around 100 pounds

While we do not have the DNA of this guy we can believe that both the Asian and African versions of this human had Y chromosome Haplotype "A" and MtDNA Haplotype mutation "L" [A_m:L_f]. This group lived a very long time on the African continent and disappeared as the mutated variant we have today took control of the country about 80 thousand years ago. At that time the "B" Haplotype and "L" DNA groups became dominate, Cro-Magnon like beings of Africa. How the Homo-Erectus base mutations were found in different locations was not by simply spreading across the world, but we will have to get to that later.

Erectus Mutation

Before we go on let me explain something important. We hear about homo-erectus mutating all over the place, but when we look at mutations of Homo Sapience, there is almost no mutations from 100 thousand years ago until 12 thousand years ago so it is unlikely that modifications of homo-erectus were "Natural". When one talks about the Neanderthal "mutation", we are told they mostly lived in Europe, but they have been found all over the place.

Mutations seem to occur during great upheavals of the Earth as the atmosphere is broken and cosmic rays can modify genes. As I mentioned earlier, from Ice Core samples, Atlantic Ocean Paleo-magnetic tests, and movement of the Hawaiian "Hot Spot" we know that the Earth went through violent changes about 350, 220, 120, and 10 thousand years ago, in fact the last turn southward of the Hawaiian chain happened about 10 thousand year ago. Not only was there nasty weather, but the Earth axis was shifted wildly. This would have disrupted the atmosphere and massive infusion of <u>radiation on the earth may have caused massive mutation of plants, animals and people.</u> We can believe massive genetic modification might have occurred around the 120 thousand year barrier but all of the hominids in this book only came into existence in the Tertiary and Pleistocene [120 to 10 thousand year spread]. There is no question that many modification did happen, but it is highly unlikely they were made without an outside influence like a genetic scientist or someone similar.

Erectus to Modern Man

We do not know the exact details of the next mutation, but Homo-Erectus [presumably the "L" mitochondrial mutation female in Africa mated with the "A" mutation male]. Some believe Erectus was mysteriously "transformed" to Cro-Magnon man as the mitochondrial mutation Female mated with the "F" mutation males somewhere south of Turkey about 40 thousand years ago <u>without transferring ANY Homo-Erectus mutations</u>. The Genesis 1, 2 and 6 account is not much comfort as it simply says.

- *[Chapter 6] Before Adam there were "giants" these were the great men of old. [During the Cretaceous]*
- [Chapter 1] *God made [Homo-Erectus] man during the 6th [Age] to* **replenish** *the Earth with people after some horrible destruction.* *[Beginning of the Tertiary Period]*
- *[Chapter 2]Afterwards he rested during the 7th age, during the 8th age, God made Adam.* [This begins the Pleistocene]

The destruction was a massive war leaving only the ANAK people. Somehow, the Homo-Erectus were made as the ANAK cried out for help. These people were not well suited for work so God made Adamic people for the Anak to govern. The book of Isaiah and Jeremiah both tell us the same thing as I presented earlier. The war at the end of the Cretaceous Period had been so bad that EVERY CITY was gone and the world was without form and void.

Jeremiah *-I beheld the Earth, all the cities were broken down.*

Isaiah—*Satan [leader of the Anak] made the world like a desert and overthrew its cities.*

The ANAK and the Titans were, essentially, the same race of people. The titans, as described by the Greeks, were simply here before the war. Most texts simply called the Titan humans, giants, but I kind of like Titan. [Think of Titan and Anak as similar.] The quick reference below tries to keep them straight.

- Titan People -Mesozoic Period only
- Anak People- 120 thousand years ago to 1500BC
- Homo-Erectus to Neanderthal- 100 to 35 thousand years ago
- Cro-Magnon- 40 thousand till present
- Cro Magnon, Anak, Homo Erectus hybrids all over the place -20 thousand till present
- End of the Pleistocene- 10 thousand years ago

I guess you can tell we are going to have to touch on Cro-Magnon in this book or we would not find any solutions, I'm afraid. First we need to find out what happened to Homo-Erectus as interpreted by the bones they left behind. When we get to the later variations or Homo-Erectus, Denisovan, and Neanderthal, we will also gain insight from DNA. Our first modification is still in Africa.

Rhodesian Man

Homo Rhodesiensis also known as Homo sapiens arcaicus is an extinct human race slightly different than Ergaster, described from the fossil remains found in southern Africa, East Africa, and North Africa. **Dating:** These remains were dated between 300,000 and 125,000 years old by nuclear dating or from **90,000 to 75,000 years ago** using newer dating methods. They certainly lived during the Pleistocene.

Brain: Cranial capacity of one skull has been estimated at 1,230 cm³.

 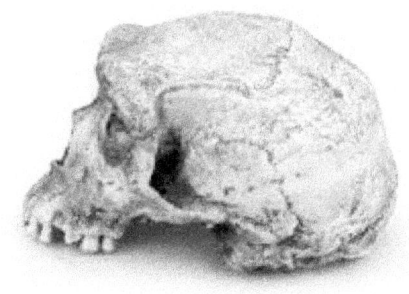

Skeleton: extremely robust individual [especially the skull], and had the comparatively largest brow-ridges of any known human as shown above

Face: It was described as having a broad face similar to Homo Neanderthalensis (large nose and thick protruding brow ridges), and has been interpreted as an "African Neanderthal" [with a much smaller brain].

Tool Making: In Africa, there is a distinct difference in the Acheulean tools made before and after 600,000 years ago by this guy as they are thinner, more symmetrical and extensively trimmed.

Teeth: Strangely, one of the skulls had cavities in ten of the upper teeth and is considered one of the oldest known occurrences of cavities. Pitting indicates significant infection before death and implies that the cause of death may have been due to dental disease infection.

SE Asia Mutations

Sangria Man

The SE Asia Variants are called Homo-Erectus rather than Ergaster. This first guy is also called Homo-Erectus Sangria. That does not mean he drank sangrias all the time, but his remains were found in Sangiran, Java, Indonesia in 1969.

 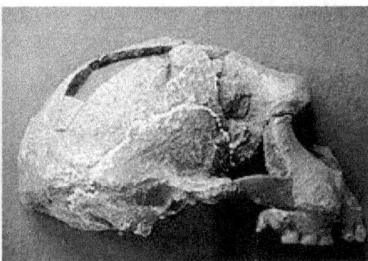 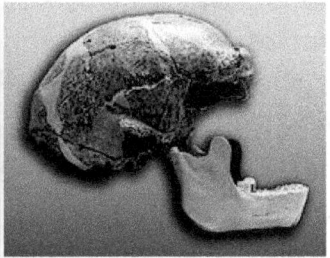

He was a cousin to the next guy also found in Trinil, Java, and Indonesian in 1891.

Java Man

Both Java and Sangria lives between 1 million and 700 thousand years ago old timing [75-65 thousand years ago new dating]

 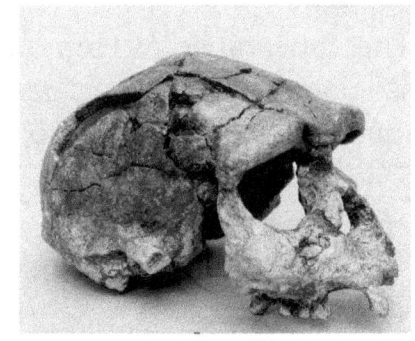

This Erectus variant was determined by the top half of an early human skull. Later it was determined that his skull was long, with a flat forehead and distinct brow-ridges and a sagittal keel [Skull protecting ridge], though many of its features have been worn flat with age.

Peking Man

Homo Erectus Pekinensis- Peking Man was a Chinese Erectus with a twist. He lived between 750 and 200 thousand years ago by old timing or about 90 thousand to 80 thousand years ago. A number of skulls were found in a single cave. Inside there was a hearth for cooking and a fireplace of sorts for heating. Not only had he controlled fire, but he set up a home. But that wasn't all. Evidence suggests he made a spear out of wood and stone. It was the first compound tool much earlier than Heidelberg man had made his version. Additionally, Peking guy drilled holes. No one knows why, but he drilled just the same and he liked soft clothing so he softened animal hides.

Tools: The appearance of Homo erectus in the fossil record is often associated with the earliest hand-axes, the first major innovation in stone tool technology.

Area of Civilization: Early fossil discoveries from Java and China ('Peking Man') comprise the classic examples of this species. Generally considered to have been the first species to have expanded beyond Africa, Homo Erectus is considered a highly variable species, spread over two continents and possibly the longest lived early human species.

Size: Ranges from 4 feet 9 inches- 6 feet

Weight: Ranges from 88 - 150 pounds

Oops-Peking Man skulls were found in 1929, but sent to the United States to keep them away from the marauding Japanese during World War II. Somehow, we lost them. The sloping forehead and thick brow ridge in front and protruding occipital torus in back are typical Homo Erectus features. Peking Man was a woodworking, fire-using, spear-hafting human race is still a mystery. Other representative skulls are shown below.

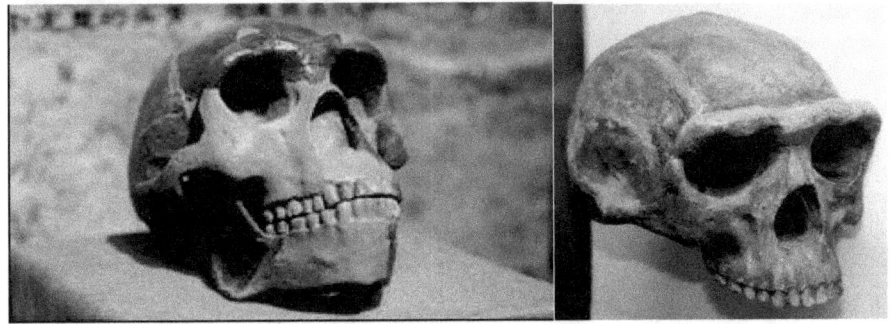

The next variant looks more modern man-like. But he had a problem.

Antecessor Erectus

Called Homo Antecessor these people lived well before Neanderthal sometime around 100 thousand years ago using new dating methods [1.2 million to 800,000 years ago with nuclear decay methods, but generally in the Pleistocene Age like all the rest]. They lived in Europe and <u>they were cannibalistic.</u>

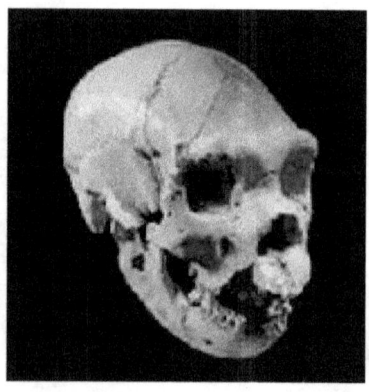

In 1994, 80 fossils of six individuals who may have belonged to the species were found in Atapuerca, Spain. At the site were numerous examples of cuts where the flesh had been flensed from the bones, which indicated that antecessor people ate antecessors. The pictures above show these guys doing just that.

Antecessor Characteristics

Size- These people were about 5½ to 6 feet tall, and males weighed about 200 pounds.

Brain- Their brain sizes were roughly 1,000–1,150 cm^3, smaller than the 1,350 cm^3 average of modern humans or Neanderthal.

Language-Based on tomography techniques it was determined that these people talked with some type of symbolic language.

Face- He had a protruding occipital bun [brow ridge], a low forehead, and a lack of a strong chin.

Where that lived- Remains have been found in Spain, UK, and France.

Tools- At one site in Spain they found approximately 200 stone tools. Stone tools including a stone carved knife. In France were found twenty tools dating back to the time of these people.

We really don't know if they walked around nude, but there were no department stores and who wants to sell clothing to a cannibal anyway?

Anak Made the Antecessor Race

There is reason to believe the Anak people converted Homo Erectus into this first modification as they tried to increase the brain capacity and make a better "servant". This seems to have been somewhat of an issue as these guys ate each other.

Idaltu Erectus

Homo Sapien Idaltu is an extinct race of humans of that lived almost 160,000 years ago in the old timing in Pleistocene Africa using the nuclear decay methods or about 55 thousand years ago with newer dating methods. As a note: "Idaltu" is from the Saho-Afar word meaning "elder" or "first born". The fossilized remains were found in Ethiopia in 1997. Three well preserved crania are accounted for, the best preserved being from an adult male having a brain capacity of 1,450 cm3. The other crania include another partial adult male and a six-year-old child. Reconstruction is shown below.

Anatomy: Considered the oldest anatomically modern humans, a couple of the skulls are shown below.

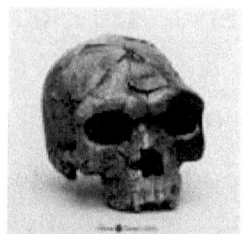
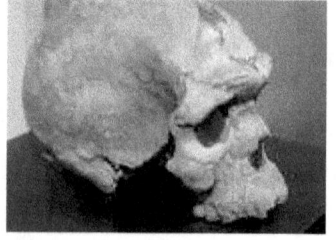

No Burial for these guys

Here is weirdness about the Idaltu people. The child's skull bore marks indicating that, after death, the muscles had been cut from the base of the skull. The rear of the cranial base was broken away and the edges polished, and the entire cranium was worn smooth as if by repeated handling. The second adult skull showed parallel scratches around the perimeter of the skull apparently made by a stone tool repeatedly drawn across the skull's surface in a pattern different from that created during defleshing, as for food. Even the nearly complete adult skull had a few cut marks. It was almost like the skull was worshipped after death. Some believe that the muscles were cut and the skull base broken to remove the brain.

Tools: many were found. More than 640 stone artifacts, and it was estimated that the area of the find contained millions of such artifacts: hand axes, flake tools, cores, flakes and rare blades.

Food: of all things, it seems these guys loved hippopotamus.

Location: The image below shows where these people lived before the end of the Pleistocene. They did not make it passed the extinction.

Heidelberg Erectus

These people were called Homo Heidelbergensis and they lived in Europe well before the end of the Pleistocene and even before the Neanderthal. They have been found with Antecessor remains so there is no telling what conflicts might have arisen. A skull is shown below along with a reconstruction of his face. This guy looked much like a modern man.

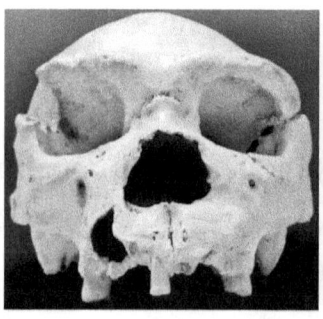  

While some have indicated that this race was the first to leave Africa, I don't think there is any reason for the assumptions except that some still want to believe that the Mitochondrial Eve DNA was the first human identified. There are no Heidelberg people who ever lived in Africa.

Some also indicate that these people might have lived as far back as 1.5 million years ago and were the first to talk, build tools, and generally be civilized. As far as we can tell, this race of people died off about 200 thousand years ago well before the time of Adam. Using the newer dating methods, this transforms to about 90 thousand to 70 thousand years ago.

Heidelberg Characteristics

Where – Heidelberg people are mostly found in Northern Spain, but they also have been found in England.

Brain size- It was about the same as modern man and slightly smaller than Neanderthal but larger than their cannibal buddies, the Antecessors. [about 1250cm^3]

Body Size – Somewhat larger than Neanderthal, 5 ½' to 6 ½', however, some have indicated that giant Heidelberg people also existed for a time.

What They Ate- We only know what this race ate Ursa Deningeri [an ancient bear] and Mimomys savini [a water vole], but they found a tibia of one of these people that had been gnawed by a large carnivore, suggesting that he had been killed by a lion or wolf or that his unburied corpse had been scavenged after death. We have not found that they had been eaten by Antecessors, but who knows.

Weapons- Hundreds of hand-axes made by these people have been found in England and Spain and 8 wooden throwing spears were found in Germany along with 16 thousand animal bones. From this German site it is believed these guys were very good hunters.

Spanish Colony- It is believed the largest group of these people lived in Spain. So far scientists have found more than 5,500 human bones dated to an age of at least 350,000 years in the Sima de los Huesos site in northern Spain. The one pit contains fossils of perhaps 32 individuals together.

Customs- We know that this race of people practiced complex funeral rites.

Anak Made the Heidelberg-Probably made from the Antecessor, These guys seemed to be much more civilized. It is not hard to believe that these were the first useful servants of the Anak people.

Neanderthals, Denisovan, and modern humans are all considered to have "descended" from Homo Heidelbergensis but there is much more to the story. Archaeological sites exist in Spain, Italy, France, England, Germany, Hungary and Greece.

Neanderthal Erectus

You probably noticed that I didn't discuss Heidelberg Haplotyping. The reason is there is little that is known. While a little more near term, Neanderthal as a race disappeared about 28 thousand years ago and came into existence about 90 thousand years ago using the new dating methods. The previous image of Neanderthal was as shown to the left, but now we understand he or she looked more like the image to the right.

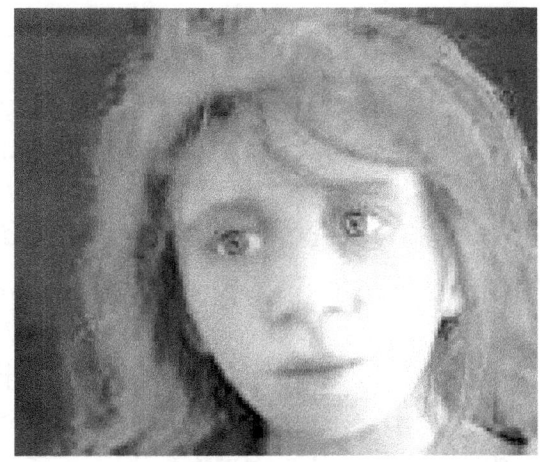

DNA sequencing- I wasn't until 2008 that enough DNA was obtained to get a pretty complete sequencing of what a Neanderthal was like. The image following shows the main locations where the remains of these people were found including 3 sites in the Middle East and one in Northern Africa. That is not to say Neanderthal was African as he has almost no similarity in DNA mutation according to the Haplotyping scientists.

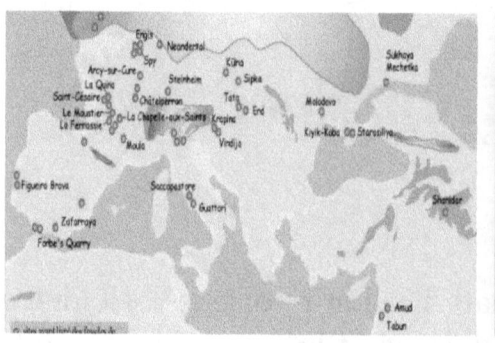 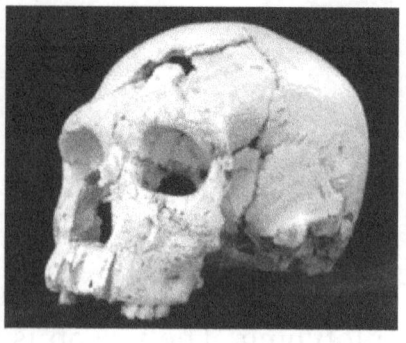

Neanderthal Controversy

Besides Neanderthal not having an African association, there are actually 2 other elements to DNA and ancestry here. The first <u>was alien DNA</u> and the second was <u>to whom the Alien DNA probably belonged</u>. An example of the robustness of Neanderthal can be seen in the image previous right. This massively thick skull is of our <u>very smart</u>, very adaptable, <u>larger brained cousin</u>- Neanderthal. Yes; he had a big nose, but he probably liked it that way. He was not a backward race, but he has caused problems when trying to understand races. Usually, the Anthropologists simply ignore him, but let's take a minute to understand his anomalous characteristics.

Wherever you find Neanderthal, you will find a variety of tools. Some are shown below. While they don't look like much there are drills, scrapers, knives, hammers, and many things we use today, but he had to make them himself as ACE hardware wasn't around.

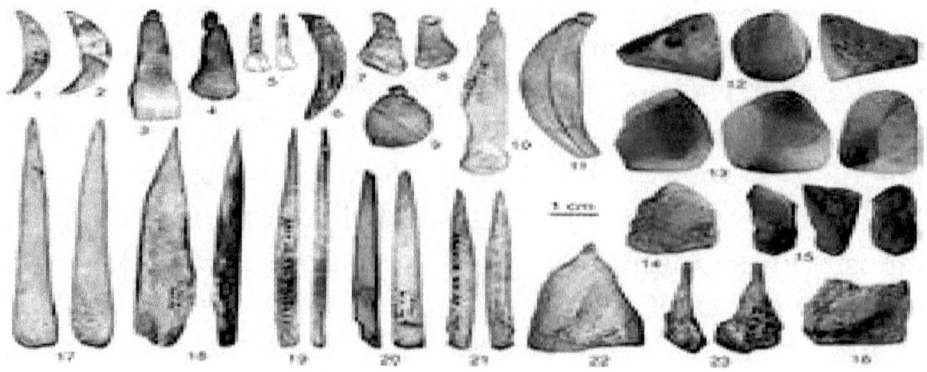

Red Talkers-Researchers tell us he possibly had a 70 word vocabulary and they carried the "ginger Gene" that makes people have red complexion and hair. Previously, Neanderthal was described as still being naked and hairy with a shaggy black beard and massive nose as shown to the left above. DNA tests demonstrated that Neanderthals possessed fair skin and reddish hair. They had a high level of "Rufosity": having reddish hair, with red pigments, or natural freckles. We can believe the Anak also had many of these traits as they were described as being red-skinned, so that part of Neanderthal probably came from them and it continued down the line to the Red Nordic races.

Tight Society-Evidence of clothing, tools, hunting skills, burials, handmade jewelry, toys, and other things allow us to know this was a well-developed and social group of people

4 Sub Races-We tend to think of Neanderthals as one species of cavemen-like creatures, but now scientists say there were actually at least three different subgroups of Neanderthals. Using computer simulations to analyze DNA sequence fragments from 12 Neanderthal fossils, researchers found that the species can be separated into three, or maybe four, distinct genetic groups as those in Western Europe, Southern Europe, Eastern Europe, and the Middle East, have DNA differences.

Art-One sign of civilization is Art. What we have found is that jewelry was made [below right], musical instruments and even the occasional carving on the wall of a cave as shown next left. While it doesn't look like much, these guys were free thinkers and creative. Additionally we found Neanderthal necklaces made from twine and shells which may show chivalry of the men providing trinkets for their lady friends--- or men that just liked jewelry.

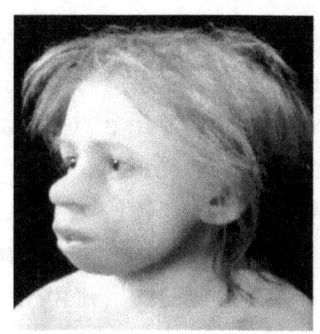  

Alien DNA-The thing to note about Neanderthal is that his DNA has some alien components. To understand it we must discuss the Annunaki people. Called the Anak by the Jewish ancestors, these special people were here before Cro-Magnon and Homo-Erectus humans and they modified DNA with their own DNA. Anyway, to make a long story short, the Anak may have lost the ability to procreate directly and they had some other issues, but they lived a long, long, long time. During the Tertiary period, they experimented and modified animals and people finally making Neanderthal well before Cro- Magnon came along, but here is the interesting thing. Researchers have found that Neanderthal DNA has key elements of Procreation missing in the X-Chromosome DNA. This suggests that they also had difficulty in having children, which possibly lead to their ending after few generations and it certainly would have frustrated the Anak scientists.

Somehow caught up in War-We are told, wars were often and bad during the Pleistocene Age. Many ancient texts describe them and some physical evidence was even found and hopefully I have given you the information you need to begin to sense what it was like living in the Pleistocene Age. As I indicated earlier, some of these "cousins" were found to have been shot during these war years. I know you were told the Cro-Magnon possibly preyed on the Neanderthal and eventually eliminated the species, but there is much more to the story.

Anak Molded Neanderthal

Anak or Annunaki [or lords of Amenti, or the Arya depending on the society describing their ancestry] were in control of the world in ancient times including the time after the Homo-Erectus had come along. This Homo Erectus was changed genetically. The following graph shows some of the changes that happened almost overnight as we compare Neanderthal and Heidelberg with Erectus. Denisovan was also added for comparison.

Neanderthal	Denisovan	Heidelberg	Erectus
Lived until -28,000BC	60,000*	65,000*	65,000*
Found "Alien" genes	Yes	Yes	??
Brain size to 1600cc	1400cc	1400cc	800cc
About 5 ½ feet tall	5 ½ ft.	5 ½ ft.	3 ½ ft.
Elongated Brain**	Similar	Similar	Rounder
Red Hair and freckles	Dark	Light	Black?
Much less hairy	Less Hair	Less Hair	Hairy
Light skin	Dark skin	Light skin	Black?
Lived in Europe	Eurasia	Europe	Africa
Made and used tools	??	Yes	Yes
Began to bury the dead	??	Yes	no
Made and used jewelry	??	Yes	no
Began to live in villages	??	Yes	no
Began to protect the sick	??	Yes	no
Began to have religion	??	Yes	no

*using newer timing   **Increased motor skills

Anak Converted Erectus Into Neanderthal

The changes certainly were massive mutations, seemingly impossible to have happen. In a twinkling of an eye, so to speak, Erectus had become Neanderthal. The results of an extensive analysis of a 50,000-year-old Homo-Neanderthalis toe bone belonging to a Homo-Neanderthalis woman, helped us understand more. Findings released by the journal *Nature*. Here are the interesting parts. For me the 1st one is of most interest.

- <u>*The Denisovan [variant of Neanderthal] share up to 8 percent of their genome with a "super archaic" [unknown species] that dates back to the Tertiary Period.*</u>
- *Ancient human species, including Neanderthals, Denisovan and Cro-Magnon mated with each other, resulting in an incredibly complex family tree.*
- *The results conflict with the "Out of Africa" theory.*
- *About 2% of the DNA of all people with European ancestry can be traced to Neanderthals and a higher percentage is found in Asians, including 6% for Australians.*
- *Brazilians and native Americans have a relatively high genetic contribution from the Denisovan/ Neanderthal.*
- *Only 87 genes responsible for making proteins in cells are different between modern humans and Neanderthals. Some of the differences involve ones involved the development of brain cells.*

The 8 percent Anak DNA tells us the Anak had something to do with the creation and development of Neanderthal. They simply were not evolutionary mutations from Homo-Erectus. Anak made them. To understand this change a little better we need only read a small section in some of the ancient texts to expand on what I already presented earlier.

Enoch 2:18-*Three [Anak] came down and copulated with women and had offspring.* [The "three" probably indicates three successive attempts at inbreeding with humans or three simultaneous inbreeding attempts.]

Nag Hammadi Creation Text-*Now come let us [the Anak people] lay hold of her [the human female] and cast our seed* *into her, so that when she becomes soiled she may not be able to ascend into the light-rather she whom she bares* [her half-breed offspring also known as Gentiles] *will be under our charge.* [The Anak had sex with or modified the Homo Erectus and later they modified Cro-Magnon humans as well. Our Bible indicates when Eve gained wisdom for the "Tree of Knowledge", she was punished by having her seed and the seed of the Anak incompatible and then she was ashamed of her genitals and covered them with a large fig leaf.]

Genesis 3:4-16- *the Nahash, Conjurer or serpent,* [One of the Anak named Samael in some texts] *said unto the woman [Eve] - God doth know that in the day ye eat thereof, then your eyes shall be opened, and ye shall be as gods [Anak], knowing good and evil.- she took of the fruit thereof, and did eat, and gave also unto her husband - and they knew that they were naked [and were ashamed of their genitals]; and they sewed fig leaves together, and made themselves aprons.--the woman said, the conjurer [Anak] seduced me- And the* LORD *God said unto the conjurer [Anak], because thou hast done this, - I will put enmity between thy seed [DNA]and her seed [DNA]- Unto the woman he said, thy desire shall be ONLY to thy husband* [This is saying Cro-Magnon [Eve] would no longer be seduced by the Anak people and even if she did, there would no longer be any offspring. It was not a restriction of the half-breed Gentiles.]

Genesis10:1-5- *Now these are the generations of the sons of Noah* [Cro-Magnon Survivors of the Pleistocene Extinction], - *The sons of Japheth; Gomer, and Magog, and Madai, and Javan, and Tubal, and Meshech, and Tiras.-<u>By these were the islands of the Gentiles divided in their lands; every one after his tongue, after their families, in their nations</u>.* [Half-breed Gentile people survived the Flood, integrated with the purebred Cro-Magnon, and had many different languages 10 thousand years ago.]

Melchizedek-*Pray for the offspring of the angels, together with seed which flowed forth from the father of all who made the entire universe from nothing there were engendered the gods [ANAK people] and angels, and <u>the men that came out of the seed,</u> all of the natures, those in the heavens and those upon the Earth—now the nature of females was wanting among those that are in the heavens. They were bound with men and women, but <u>these [gentiles] were not the true Adam</u> nor the true Eve.* [This verse talks about a difference between angels and people called the ANAK and infers that a union between man and one of Anak was accomplished. It specifically indicates this [Half-breed or Gentile human was not the true Adam and Eve.]

Middle Eastern Changes recorded- As I brought out earlier, the Sumerian, Indian, Iranian, South American, North American, and just about everywhere else had similar descriptions of genetic manipulation. *"After having sex with one of the Annunaki , the man was no longer to be associated with beasts. <u>His hair was no longer all over his body</u> and his legs were diminished. He became a hybrid.* [I think we will someday find that Homo-Erectus had hair all over its body while we know Neanderthal had a light complexion and red hair.]

Greek Homo Erectus Change-The Greek Mythology was not as specific, but the Greeks did indicate that the early attempts at gods having offspring with man resulted in monsters. Here is what the Greeks said.

"First only existed chaos and Gaia which would be the Earth was formed. Gaia gave birth to the sky. The gods Gaia and Uranus had children-12 were Titans, 3 were Cyclopes <u>and three were monsters with 100 hands. One of the titans took control [Cronos] and created the gods</u> who took control by freeing the monsters. Zeus had a woman created from clay [Pandora]. She opened the box of knowledge and released evil

into the world." [Later Zeus became a lecher with human women. Anak and gentile humans mated without restriction.]

The Anak had converted Homo Erectus into the Neanderthal. The hairy, miniature, small brained man had changed completely. Like those before, we can believe the Anak Race was responsible for the "update" in Heidelberg people. With more creativity came more capability. As far as we know, there was no major upheaval that forced massive DNA mutation as we see after the worldwide flood. Somehow a new race of humans came along and with them there was mystery. Luckily we found DNA. This helped some questions to be answered, but many are just as open today. Before we go on, let me make something clear.

Gentile is not a non-Jew

Some in the religious community try to make the word Gentile mean non-Jewish, but that is not what the Bible tells us as the Homo-Erectus half breed Cro-Magnons were gentiles.

Genesis 10:5- *By the sons of Japheth were the isles of the Gentiles divided in their lands; every one after his tongue, after their families, in their nations.* This was immediately after the flood showing that gentiles survived and that they were here well before there was such a thing as a Jew.

Testing Neanderthal DNA

As scientists started looking at DNA from a number of sources we started to get a more definitive accounting of what Neanderthalis was like. As I mentioned, we now believe Neanderthal were light skinned and red-haired.

Red-Headed Neanderthals-Ancient DNA has been used to show aspects of Homo-Neanderthalis appearance. A fragment of the gene for the melanocortin 1 receptor (MRC1) was sequenced using DNA from two Homo-Neanderthalis specimens from Spain and Italy in 2007. Neanderthals had a mutation in this receptor gene that **has not been found in modern humans**. The mutation changes an amino acid, making the resulting protein less efficient. Modern humans have other MCR1 variants that are also less active resulting in red hair and pale skin. The less active Homo-Neanderthalis mutation probably also resulted in red hair and pale skin, as in modern humans. Because this (MRC1) stuff was in Neanderthal and not in us, this gene came from someone else. Many scientists struggle with this because they will not accept the existence of the Anak no matter how much evidence shows up.

Speaking- Another odd sequence called [FOXP2] was found. The FOXP2 gene is involved in speech and language and this FOXP2 gene is mutated in Chimpanzee, but not in Neanderthal. Therefore it is believed Neanderthal did not just grunt when he saw a good looking woman. He might have said something about her eyes or shape rather than clubbing her in the head as we once believed, but recent physiological discoveries indicate that their voices were high pitched and nasal, not the baritone grunts we normally associate with cavemen.

Blood type- Interestingly, researchers found some Neanderthalis people had type O blood [possibly OA or OB]

Microcephalin gene –This one has been in the news lately as the news media claimed Zika virus would shrink the heads of babies making them Microcephalin. Researchers have determined that this gene comes from the Haplogroup D. Initially it was believed Neanderthal carried this problem gene, but later studies disproved it. One can still believe Microcephalin genes came from a Homo-Erectus that was not Neanderthal.

Comparison to Chimpanzee- As I showed earlier, researchers compared the Neanderthalis to modern human and chimpanzee sequences. Modern human sequences varied between each other with about 8 substitutions in the DNA chain, but chimpanzee sequences by about 55 substitutions when compared to the mean of modern humans and Neanderthal had about 27 substitution, so by this 1997 study and another in 2000 we could say Neanderthal was half way between Chimpanzee and Modern man.

When Did a Differenced happen?- Researchers found modern human and Neanderthal human remains from about the same time frame and found that the modern looking human had modern DNA sequencing while the Neanderthal had sequencing similar to the other Neanderthal. This showed that early anatomically modern Homo sapiens were not very different genetically from current modern humans, but were still different from Neanderthals. It was as if Cro-Magnon was a new species that simply appeared and didn't change much since it "appeared".

Europeans and Neanderthal- Many Neanderthal remains have been found in Europe and European skulls are slightly more Neanderthal-like that other parts of the world so researchers believed that would finds a link but there has been found no closeness of Europeans than any other modern human. Also various analyses have examined the amount of Homo-Neanderthalis contribution to modern human mtDNA. The analyzers were unable to find positive evidence for interbreeding between modern and Neanderthal humans. It was as if the group that became Europeans, settled there after the Extinction of the Pleistocene Age and had nothing to do with who had been there before. This is an important point as many try to force fit colonization by location of ancient remains from before the Pleistocene event. No wonder there is so much confusion, I will discuss this more later.

A 2002 study concluded- *Homo-Neanderthalis and modern human mtDNA is consistent with large-scale replacement and some small amount of interbreeding between modern and archaic[Erectus] populations. Interbreeding between archaic and moderns may have involved different species of archaic humans, including populations in Africa, Asia and Europe.* It showed more similarities between non-African modern humans and Neanderthals than between African modern humans and Neanderthals but all differences are very minor. Approximately 2.5% of non-African modern human DNA is shared with Neanderthals while Africans share only 1.5%.

No change- A study in 2009 indicated that Neanderthal changed only about 1/3 as much as the changes noted in Modern human DNA. This doesn't make sense in that Neanderthal supposedly lived for a longer time period. If researchers would simply get off of this nuclear decay timing, they could understand that modern humans have been around longer than Neanderthal. Using nuclear decay timing it was determined that the spit that resulted in modern humans from

Neanderthal occurred about 450 thousand years ago, and their heads were about to explode. One would expect with modern humans being on Earth 40 thousand years and Neanderthal living over a period in excess of 400 thousand years. Using the newer dating methods, Neanderthal only existed about 50 thousand years total so the tiny number of changes makes a lot more sense.

Neanderthal location Mystery-One mystery seems to be how people got where they were. Neanderthalis DNA contribution has been very scarcely found in African populations, but there is an exception when testing the non-African DNA portions of the Maasai, in East Africa. Not only is there enough similarity to show contact, DNA-ologists can tell when the contact occurred. It can be concluded that recent non-African gene flow was the source of the contribution as about an estimated 30% of the Maasai genome from about 100 generations ago. [2500 years ago- not 100 thousand years ago as some had speculated.] All that being said; a new human type showed up before the time of the Cro-Magnon rise to power. It was good old Neanderthal. It just appeared as if a scientist modified DNA to augment the Homo-Erectus people.

Complete Sequencing of Neanderthal DNA

Let's look at the DNA charts from before in a little more detail. They seemed to show how very close Neanderthal and modern humans were. This first graph developed in 1999 by Dr. Krings and his DNA scientists shows distributions of sequence <u>differences</u> among Humans, the Neanderthal, and Chimpanzees. This is what they stated. *It may be noted that a small fraction (0.04%) of the inter-human comparisons are larger than the smallest distance (29 substitutions) between the Neanderthal and humans.* In this initial work seemed to show Neanderthal could be considered a modern human.

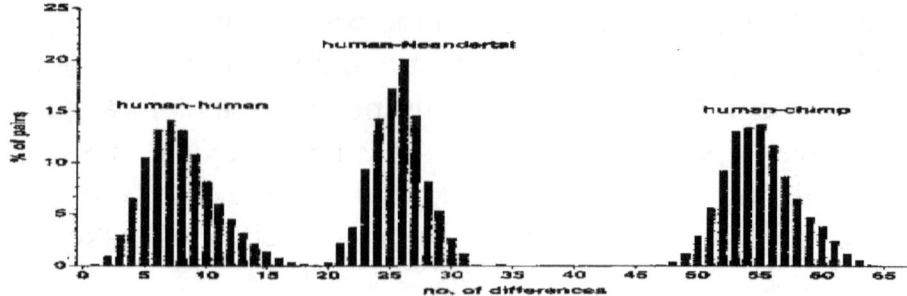

As I said it was found that this was not a correct assumption. In 2008 A team under Dr. Green were able to sequence almost all of the Neanderthal DNA. The chart below is from his study which not only looked at numbers of differences, but also where the traits were positioned in the DNA sequences. It not only shows there are 2 types of modern humans, but also that there was a wider separation than originally was believed. Neanderthal had very small differences to modern humans and much closer than chimpanzee.

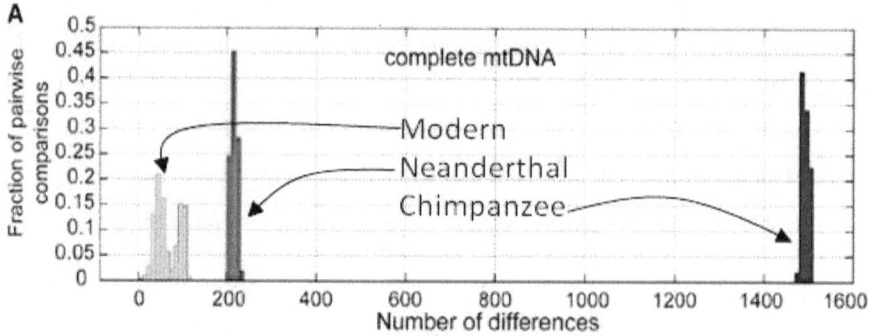

In May 2010, Dr. Green provided us the first Neanderthal nuclear DNA scan of about 2/3rds of the entire thing from 3 specimens which show that Neanderthals interbred with humans, and that all non-African modern humans contain 2.5% of Neanderthal genes. They also suggested that because of the high levels of Neanderthal traits in Asians that Neanderthal probably mixed with Cro-Magnon in the Middle East as almost NO Neanderthal characteristics are found in African peoples.

Where are the People with Neanderthal Genes?

While we found most Neanderthal remains in Europe, testing modern humans describes something strange. The following map shows where people live that have more Neanderthal DNA traits that the rest of us. This includes large tracks of North, Central and South America, the Far East and Oceana. [I circled the areas of concentration so they would be easier to see.] There are almost no people with Neanderthal traits in Africa and Australia, and few in Europe.

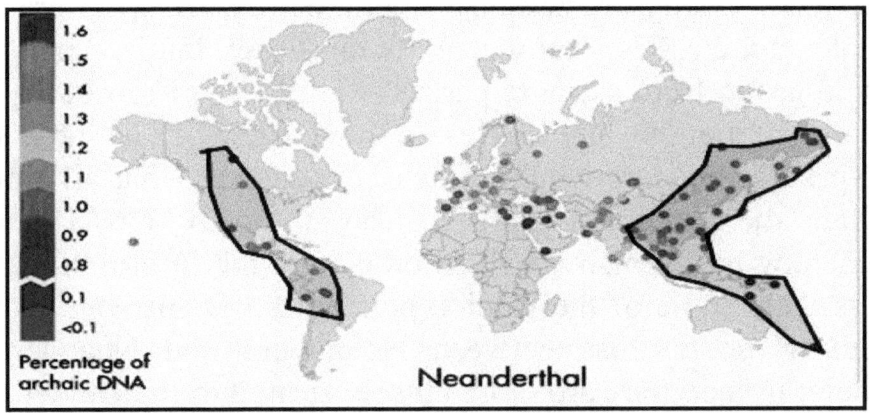

The reasons for a reduction in Neanderthal DNA traits in Europeans and the lack of African Neanderthal traits caused researchers trouble because they refused to use ancient texts to help them understand. I spent some time on Neanderthal as he almost seemed to be the highest evolved man before Cro-Magnon came along, but then we found Denisovan. Similar to Neanderthal, this race of people had some anomalies.

Denisovan Erectus

Let me let you in on a dirty secret. If you want to produce and sustain a viable community, you must have <u>many opportunities for procreation</u>. Some don't work, sometimes location is a huge deterrent, sometimes the offspring are not viable, etc. etc. That being said, tracing back family lines to a single group from a single parent is not likely. I'm not saying there will not be similarities in how each of the potential offspring producers would mutate, but to think that all would mutate the same is--- OK I'll say it again; UNLIKELY. Somehow the Denisovan Race didn't get the information and messed up some of the Haplotype studies. Denisovan people seem to be a cross between Heidelberg and Neanderthal people; in fact, they are more closely related to the Heidelberg according to their MtDNA. Recreations are shown below.

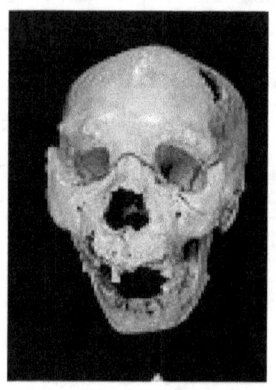  

A higher quality Denisovan genome published in 2012 revealed variants of genes in humans that are associated with **<u>dark skin, brown hair and brown eyes</u>** are contained in this race showing he looked substantially different than Neanderthal [Last image above]. While researchers didn't get a substantial amount of the DNA, they did get enough to provide a snapshot of the differences between Neanderthal and Denisovan. Closer than Chimpanzee, Denisovan was still much more removed from modern man. While the first Denisovan bone was found in Russia, the more interesting thing is that this seemingly close relative to Heidelberg and Neanderthal who both were European had its highest concentrations of similarity to the Australian Aborigines.

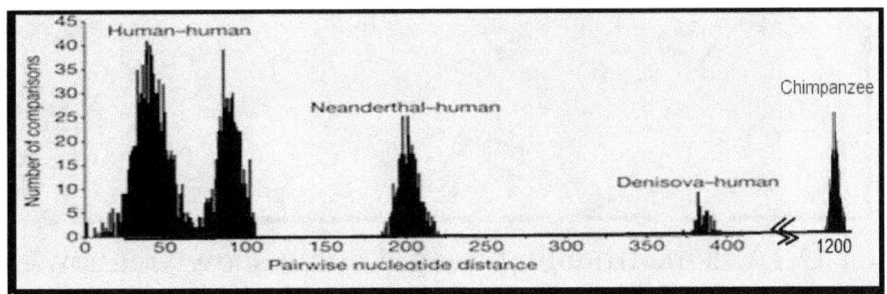

Today we find very few Denisovan specific traits, but where we find them may be telling as there are none in Africa, North America, and northern Asia. Don't worry about this map too much as the maximum similarity of Denisovan is about 0.5%.

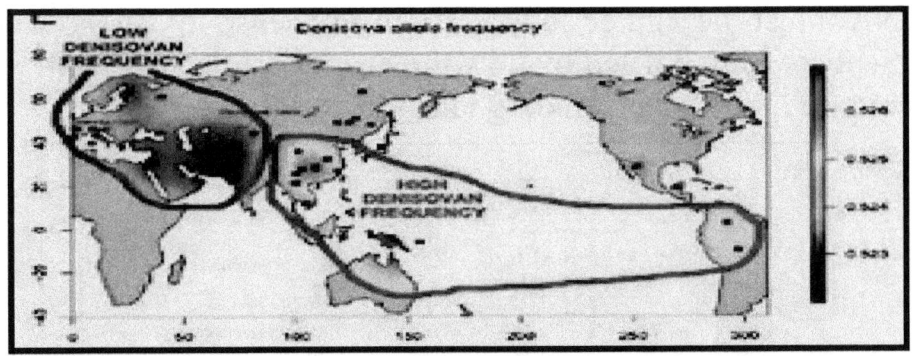

Not Out of Africa-I know you keep hearing a new mutation of people came out of Africa 200 thousand years ago, but it simply didn't happen or the probabilities are extremely low except for the Homo Erectus people. Certainly, there were early people in that country, but there were other people in other locations as I have been presenting. One race of people was found in the Far East and SOMEHOW these guys mated with a European Heidelberg from Spain. The map below shows where a finger bone of Denisovan was found in Russia.

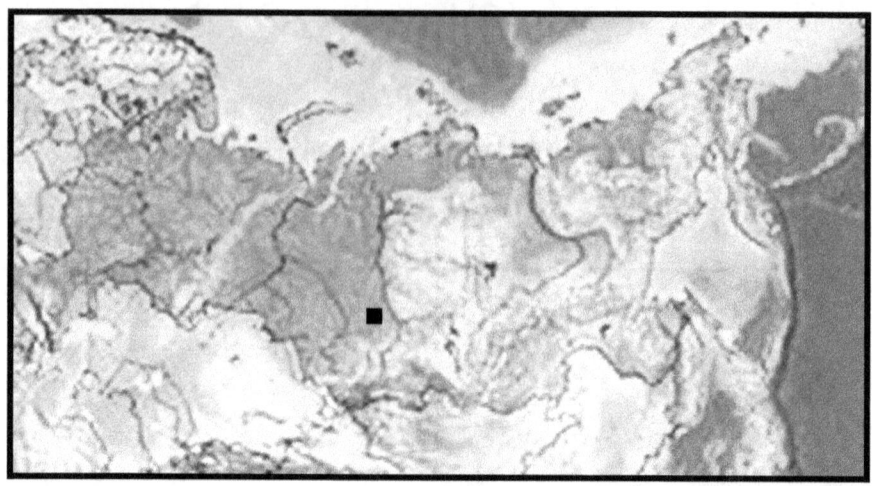

Their DNA is the troubling thing. Let me show you how they try to resolve DNA connections. In the following graphic, you will see Denisovan are genetically similar to Melanesians [Australians and New Guineans] and that is simply not good because Melanesians are directly related to the various Cro-Magnon races. Also, notice that Neanderthal and Heidelberg DNA must be linked oddly to make sense of it as well. Scientists were baffled by how this was possible. We will put sanity into our understanding later.

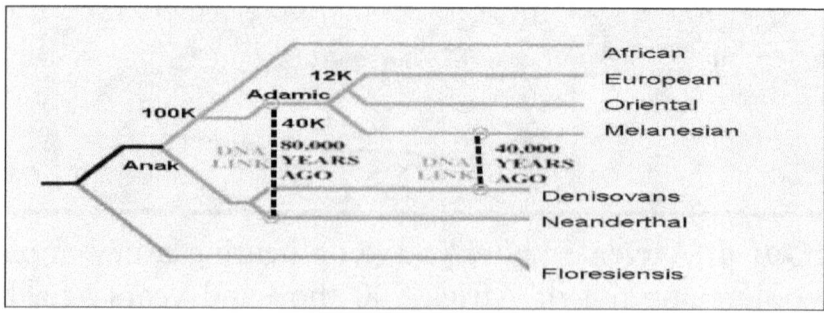

Denisovan Mutation-Denisovan were "Generally" Neanderthal that were not located with Neanderthal. Completely isolated from the REAL Neanderthal, some of their DNA changed during a mutation period differently than the European version.

Wrong Denisovan Tracking-An international group of scientists have completed a highly detailed analysis of DNA from what is estimated to be a 50-80,000 year old Denisovan finger bone. Common thinking from anthropologists without insight now say about 700,000 years ago [about 90 thousand years ago by new dating], a group of humans left Africa and spread out across Europe and Central Asia. Those walking the thousands of miles to the Far East and those walking to Europe stayed the same somehow [their DNA did not mutate]. Here is where we must say there are no Neanderthals in Africa so as these people left, both groups were simultaneously zapped with cosmic rays or something and both became what scientists call Homo-Sapien-Neanderthalis. During the thousand years of wandering to get to Cambodia and jump over to Australia, there were NO significant changes of either group and the Africans never had a cosmic blast to make any similar people even though there such a HUGE likelihood of the change from Erectus to Neanderthal that 2 completely separate groups had the SAME mutation. Oops! I laughed a little just then, but I'm over it now. This is what happens when a scientist gets so involved with his particular insight and disregards all other information around them. No one even thought to ask how the finger bone got to Siberia.

Besides this guy that must have taken a wrong turn to go North, soon they go tired of walking and had kids. The kids of both groups stayed the same. Both are Neanderthal except for the unusual DNA that somehow was introduced. These people became Neanderthals and Denisovan, and the people they left behind in Africa or the Middle East became Homo-Sapien-Sapien. So how does a Denisovan finger bone illuminate what happened to Neanderthals? One thing that we now know is there are 23 known areas of the genome that modern humans do not share with either Neanderthal or Denisovan.

Flying-Another thing that is almost certain is that flying transports were common in the olden times allowing for widely spaced separation of individual, societies, and trade. Hundreds of document, physical evidence, and artwork describe them, explain how they were used and allow us to understand more about DNA if someone would simply try to explain something in a less comical way.

Cross-breeding-Scientists have now found a crossbred individual "half Neanderthal and half Modern man". When the DNA structure of Europeans is examined, it is found that they now are between 1 and 4 percent Neanderthal. When I say cross breeding, I have to also add in Denisovan.

Oldest DNA-In an underground cave in the Atapuerca Mountains in northern Spain, a pile of bones was found. In the pile was a leg bone with DNA. The bone is apparently 400,000 years old [old dating] or about 50 thousand years old [new dating] and shown below.

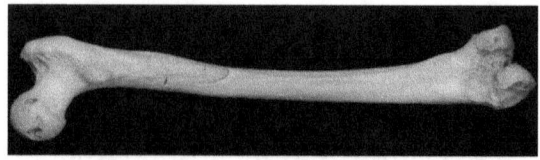

The researchers reconstructed a nearly complete genome of this fossil's mitochondria. The fossils unearthed at the site resembled Neanderthals, so researchers expected this mitochondrial DNA to be Neanderthal but to their dismay, the ancient Mother was Denisovan. Now it seems that these Denisovan stayed with Neanderthal and went northwest to get to Spain. Then the Denisovan group was mutated without the Neanderthal getting exactly the same mutation and most of the Denisovan left for Australia while at least the finger of one of them went north to Siberia.

Please don't get me wrong. I don't believe this story. It is what they are trying to make everyone else believe. From this, let's go to something we have details about. We know quite a bit about the beginning of the "8th Age" and the "creation of Adamic people". Scientists call them Cro-Magnon or Homo-Sapien-Sapien. Don't ask me why Sapien has to be written twice.

Inbreeding-In 2010, it was concluded that the Denisovan population shared a common branch with Neanderthals or Heidelberg from the lineage leading to modern African humans. This suggested that the divergence of the Denisovan mtDNA resulted from the persistence of a lineage purged from the other branches of humanity. A detailed comparison of the Denisovan, Heidelberg, Neanderthal, and human genomes has revealed evidence of a complex web of interbreeding among the lineages. Through this interbreeding, 17% of the Denisovan genome represents DNA from Neanderthal population, while evidence was also found of a contribution to the Denisovan from an ancient human lineage yet to be identified. I'm not saying this "alien" DNA was Anak DNA-----Maybe I was!

Artificial Mutation-For the average person about 4% of your DNA is known to be from Neanderthal; whereas, about 5% of the genome of Melanesians DNA came from Denisovan. Tibetans have a DNA Haplotype mutation that assists with adaptation to low oxygen levels at high altitude and in 2014 it was found that Denisovan had the same mutation showing their close relation. With the massive distances between each of those that seemed to be connected with Denisovan, it is easy to conclude that the Anak were working to modify and make Neanderthal more rigorous. One group of Anak must have tried their genetic art in Indonesia.

Floresienesis Erectus

As Homo Erectus wandered in the Orient, there was a mutation that occurred about 70 thousand years ago [using new dating system]. This new mutation was called Homo Floresienesis (nicknamed 'Hobbit'), have been found between 50,000 and 17,000 years ago on the Island of Flores, Indonesia.

Size: 3 ½ feet tall standing with shrugged-forward shoulders

Weight: 30 kg (66 pounds) - estimate from a female skeleton

Teeth: large teeth for their small size,

Chin and Face: no chins, receding foreheads, tiny brain

Feet and Legs: relatively large feet due to their short legs.

The images following first show Floresienesis skull compared to a modern skull and the second shows that even with most of the animals on the island being miniaturized, they were still much larger than Floresienesis.

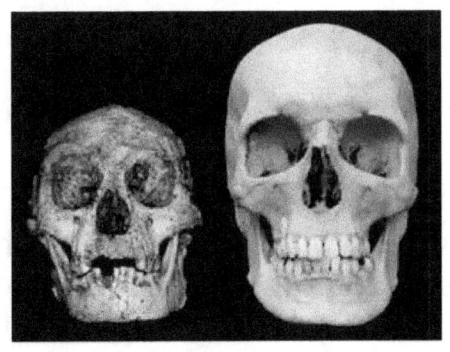 

Tool Making: Despite their small body and brain size, they made and used stone tools,

Occupation/food source: hunted small elephants and large rodents, coped with predators such as giant Komodo dragons.

Fire: may have used fire.

Haplotyping: from limited DNA details, Floresienesis are blocked as LB-1 Haplotype. Separating him from Homo Erectus.

Why Were They Small?-It seems everything on the Island got small. Potentially some external mutational element existed here, but it is not clear. Pygmy elephants on Flores, now extinct, showed the same adaptation. The smallest known species of Homo and Stegodon elephant are both found on the island of Flores, Indonesia.

End of the Race-As Floresienesis did not go north towards the Cro Magnon fighters, they survived when the Neanderthal died out. It is apparent that they did not survive the extinction period at the end of the Pleistocene Age but it probably was more about their tiny size. Unfortunately for other races, the Cro-Magnon came along and may have been responsible for killing off a number of the ancient people. Let's look at Grimaldi.

Grimaldi Erectus

Cro-Magnon's struggles did not end with Neanderthal. It has been shown by mtDNA traces that afterwards, African races entered Europe around 40 thousand years ago, specifically, and the Congid [pre-Negriod race] tried to take their piece of Europe. As the pre-Negriod "Grimaldi-Man" shown below lived, fought, and eventually died off in southern Europe between 40 thousand and 10 thousand years ago. As the Cro Magnon was advancing, it seems this group outlived the Neanderthal. Grimaldi was established as the Cro-Magnon and Neanderthal had mixed with Homo-Erectus to establish this briefly lived race.

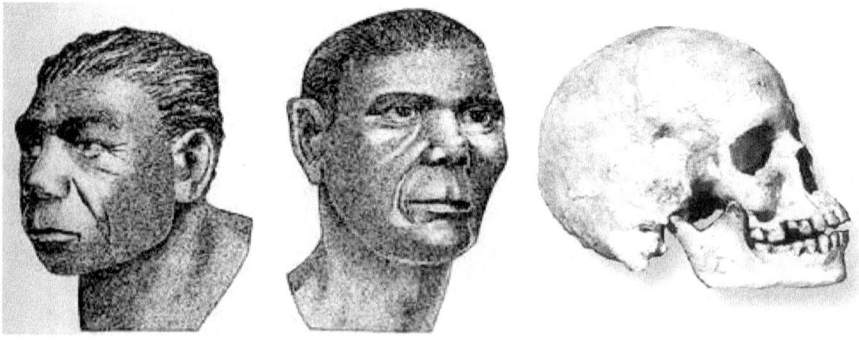

In the early 20th century, two Paleolithic skeletons were found. The skeletons differ markedly from the contemporary Cro-Magnon finds from other parts of Europe, the Grimaldi find was originally classified as a Cro-Magnon offshoot, and the features were substantially Negroid. One of the two skeletons belonged to a woman past 50, the other an adolescent boy.

Grimaldi-Man Characteristics-Build: The Grimaldi skeletons were somewhat slender and gracile, even more so than the Cro-Magnon finds from the same cave system.

Size: The Grimaldi people were small. While an adult Cro-Magnon generally stood about 180 cm tall neither of the two skeletons stood over 160 cm [5 ' 2"]. The boy was smallest at a mere 155 cm.

Brain: The skulls of the two had rather tall braincases, unlike the long, low skulls found in Neanderthals and to a lesser extent in Cro-Magnons. The cranial capacity was also quite large for their size [about 1580cc].

Nose: The faces had wide nasal openings and lacked the rectangular orbital and broad face so characteristic of Cro-Magnons. The nasal bones gave a high nasal bridge, like that of Cro-Magnons and modern Europeans and very unlike more tropical groups.

Muscles: Total muscle mass determined showed the two would have been well muscled in life, rather than having the slender build usually seen in tropical people

Family: As shown below, family members were buried together to show their closeness, family unity, and close-knit society.

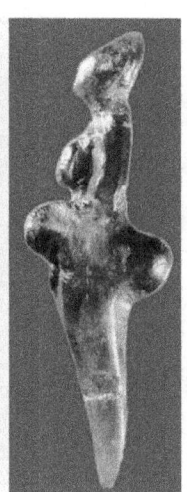
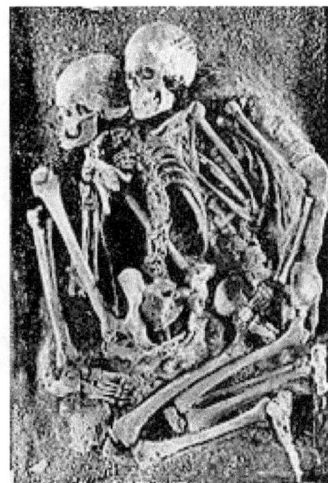

Art: Fine artwork has been found in the shape of females with a large butt as shown. Additionally caves occupied by these people. One such artwork is shown below.

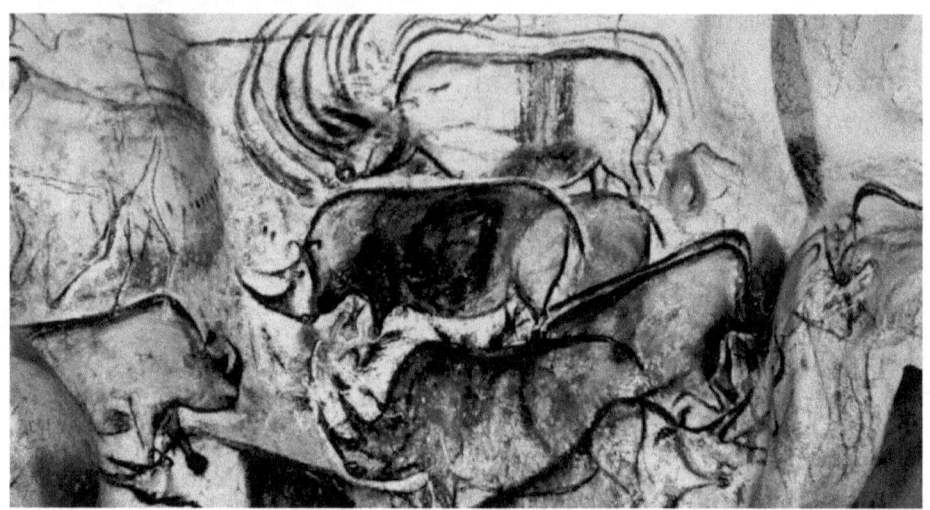

Boskop Erectus

Homo-Erectus didn't just spread into Italy. Scientist found some strange people that lived during the Pleistocene that had huge brains. Known as the Boskop Man [**Homo-Capensis**] because he was found in Boskop, South Africa, these people must have been "worked" on by the Anak scientists. Reported to have a brain capacity of as much a 30% over a modern human, this guy must have been something. He was certainly pre-Negriod, but no one could mistake his brain. Boskop man lived in southern Africa between 30,000 and 10,000 years ago and he evidently did not survive the flood at the end of the Pleistocene. However, similar skull structures have been noted in modern Bushmen or Hottentot people. A comparison of a Boskop Man and modern skull is shown below. The large Boskop is to the left.

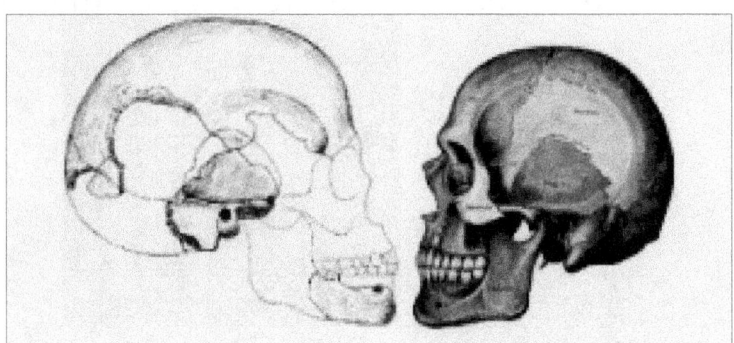

Body Size: The Boskop people were average sized

Skull: Dolichocephalic and 25% larger than modern man. The skull is unusually thick so that the brain size may only have been 1700 to 1800cm.

Brain dimension: Like the Grimaldi brain, it was more of a rounded brain than the longer brain of Cro Magnon.

Nose: Similar to Grimaldi, their faces had wide nasal openings and lacked the rectangular orbital and broad face so characteristic of Cro-Magnons. The nasal bones gave a high nasal bridge, like that of Cro-Magnons and modern Europeans and very unlike more tropical groups. Another image of the skull is shown below. Notice the teeth protrude more and the chin is much less pronounced when compared to Cro-Magnon.

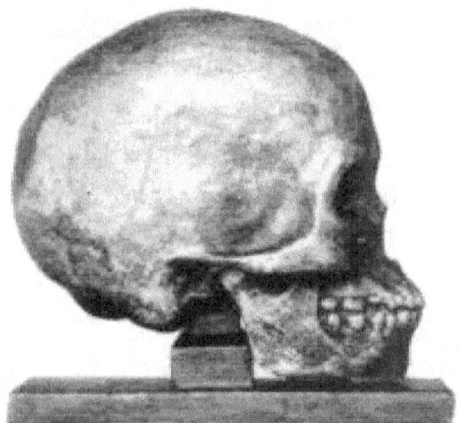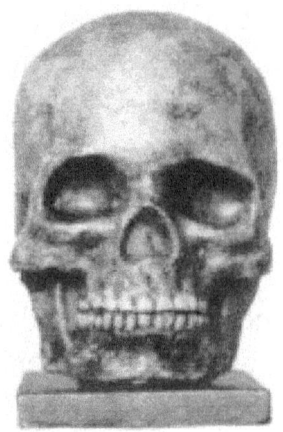

It is believed this race was a cross between Anak and normal sized human and that brings us to Cro-Magnon who is coming up next.

Cro-Magnon Race [Fm:Nf]

To show DNA mutation Haplotype we can say Cro-Magnon has the male [Y chromosome DNA mutation "F" and the Female [mitochondrial DNA mutation "N"]. Clearly the Cro-Magnon, Homo Sapien Sapien, identified by some anthropological experts as the _**most "evolved" human type ever to have lived on Earth**_ came about 40 thousand years ago. Most have no idea how he evolved so quickly. To make it stranger, there were no massive extinctions or astronomical disasters that would have increased mutation when he came along.

He was just there as shown above Y-Haplotype F and MtDNA N. Both brand new distinctions. From this group came the Armenians, White Nordics, and Red Nordics that were the cornerstones of racial distribution around the world. The images of skulls of Cro-Magnon are shown next. Many, due to his unbelievable and sudden beginning also refer to this human as the Adamic man. [Sort of the first Jew, Adam was known to be different that the others in the world at that time. The Homo-Erectus descendants in Africa, Neanderthal in

Europe, Heidelberg people, Antecessor of the UK, Denisovan, and others all would essentially be replaced by this guy or be crossbred.

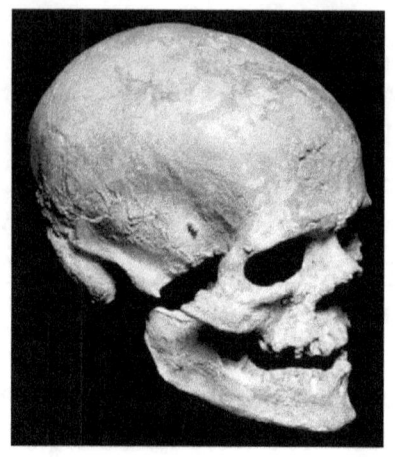 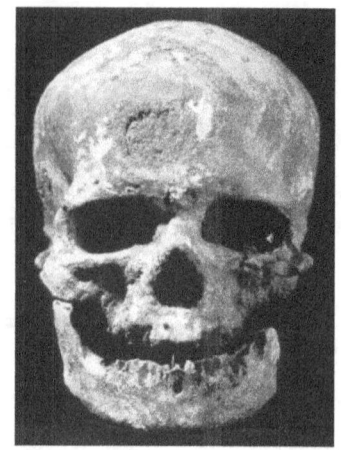

As shown below the northern half of Europe was covered in Ice for much of the Pleistocene as one group of the Cro-Magnon left the Middle Eastern area and went into the Mediterranean area after mutating to $[I/J_m:K/N_f]$ to become the white Nordic Race. Another went towards the Armenian area after mutating to $[G_m:N_f]$ to become the Armenian Race. Another group went farther north after mutating to $[R_m: H/V_f]$ to become the Red Nordic race. Another group went south after mutating to $[J_m:M_f]$ to become the Moslem and Khemetian Races. Still another group went west after mutating to $[K/O_m:F/Z_f]$ to become the Oriental Race. These and other mutations occurred at the end of the Pleistocene Age.

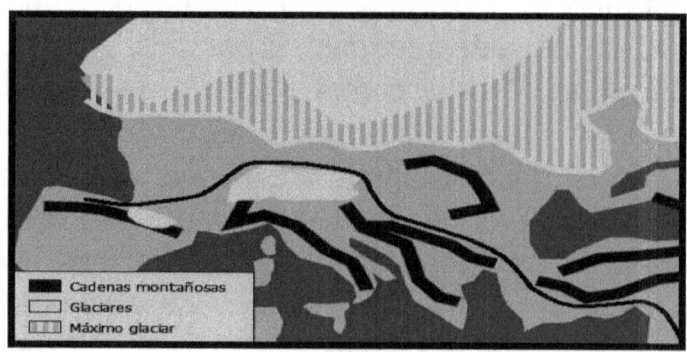

Cro-Magnon Characteristics

As for Cro-Magnons, they're pretty much just like the known European example of our species—living between 35,000 and 10,000 years ago—and are actually modern in every anatomical respect. They did, however, have somewhat broader faces, a bit more muscle, and a larger brain. So how'd they utilize their large brains? Cro-Magnon man used tools, spoke and <u>probably sang</u>, made weapons, lived in huts, wove cloth, wore skins, made jewelry, used burial rituals, made cave paintings, and even came up with a calendar. Specimens have since been found outside Europe, including in the Middle East.

Brainy Cro-Magnon

Ancient historical records of the Ulga Mongulala, the Maya, and Judeo-Christian sects tell us this man could communicate with others without using his voice. He invented and used flying machines, levitated objects, healed people by touch, went into space, and many things before the Bharata War messed up his brain and it atrophied to the size it is today. Here are just a few of the many texts describing Cro-Magnon.

__The Jewish Essene-Jubilees 10: 24-26__-And he confounded their language, and they no longer understood one another's speech, and they ceased then to build the city and the tower. For this reason the whole land of Shinar is called Babel, because the Lord did there <u>confound all the language of the children of men</u>, and from thence they were dispersed into <u>their cities, each according to his language and his nation</u>. [By saying it three times the writer is pretty sure, some limitation in man's capabilities occurred 5000 years ago and it happened to everyone on the earth, not just the tower builders. It happened during a terrible war.]

__The PreMaya- Popul Vuh-__ They [the original Cro Magnon descendents from Aztlan], <u>had the power of understanding;</u> They saw and could <u>immediately see far</u> [really good vision.]

They succeeded in knowing everything that could be seen or known in the world. Things that were hidden in the distance they could see without moving first. [out-of-body movement or television?]Their wisdom was great; they controlled the forests, the rocks, the lakes, the seas and the valleys. [control over nature.]They investigated the four corners of the earth, [Fast Transport] They investigated the four corners of heaven [space travel]They investigated the round surface of the earth. [circumnavigated the globe]

Then one day the heart of heaven blew fog in their eyes. They could not see clearly any more, like breathing on a mirror. Their eyes were covered and they could only see things that were nearby. This was the way that the wisdom and knowledge of these first people was destroyed.

From the Brazilian Ulga Mongulala Secret History-*The gods taught us the secret of man, animals, and plants.* [These were the Anak people teaching the Cro-Magnon Descendants]-*The Blood Age was the beginning of the Mongulala history. It started immediately after the Golden Age, about 10,500BC. There was an island in the west and a gigantic mass of land* [Cultural Island] *in the northern part of the ocean. Both lands were buried under an enormous tidal wave during the first Great Catastrophe. It occurred towards the end of the war between the two divine races. The war between the two divine races did not only lay waste to the earth, but also <u>to the worlds of Mars and Venus</u>.* [Space Travel and colonization] *The secret documents are kept in the underground Great Temple of the Sun. The documents testify to the 11,000 years of history and are kept in a room which is hewn out of the rock. Here also are the mysterious drawings of our Ancient Fathers. The documents are also engraved in green and blue on a material unknown to us. Neither water nor fire can destroy it. Documents were left by the Akamim gods [the Anak people], which have remained hidden. They tell about the* *matter from which everything is made. They tell about the course of the stars and the relationships in nature. The ancient priests explored the spiritual forces of man. Our priests* [Cro-Magnon Descendants] *have learned how to make <u>objects fly through space.</u> They learned how to <u>open the body of the sick without touching it.</u> They had <u>ships faster than birds' flight</u>, ships that reached their goal without sails or oars. They know how to <u>transfer thought without words</u> over the greatest distances. By magic, the ancient ones <u>suspended the heaviest stones, flung lighting and melted rocks</u>.*

I know scientists have indicated widespread inter marriage and cross breeding, but when one researcher was asked, his answer seems more reasonable. He said not likely--

- They were probably physically repulsive to each other
- They couldn't meaningfully communicate
- Beer wasn't invented yet.

Features

Skull: dolichocephalism, a not totally vertical forehead, thin and long skull with a squared jaw, sharp chin and large straight nose with low bridge. Brain size increased to the highest level yet at about $1500 cm^3$.

Size: tall stature 6 to 8 feet

Face: broad face when compared to modern White Nordics with perfect dentures.

Muscularity: stronger skeletal consistency, a higher muscular development than White Nordic

Occupation: Hunter gatherer. The hunter culture showed up with spearheads, arrowheads, richly decorated spear throwers, harpoons, and assegais, cave paintings filled with hunting scenes, whistles, and horse's head figures. Later weaponry was mixed with those of the Anak that we will investigate later.

Activities: Living a life of violent and constant physical activity outdoors. This developed them as an athletic, graceful, gymnastic human type, and when the climatic conditions got milder, this probably became more apparent. It is in the Cro-Magnon communities of Spain, France and the Balkans, where we have to look for the origin of Greek athletic traditions like races, javelin throwing, hunting, fighting and archery.

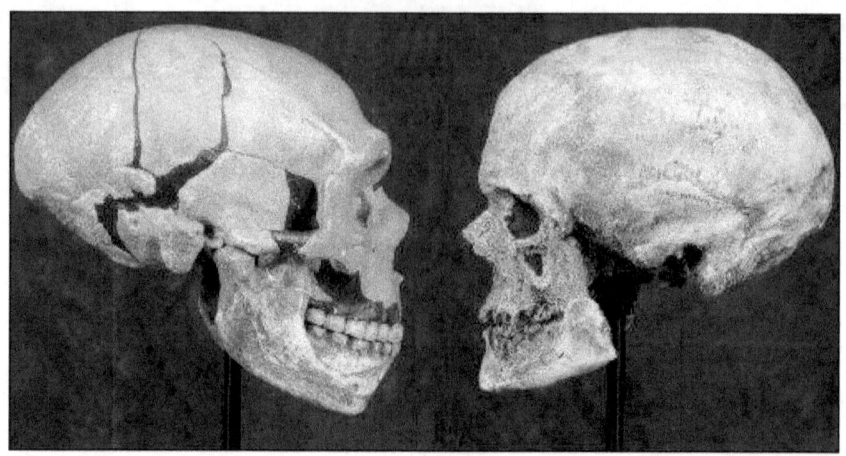

The Neanderthal (left) had been in Europe for over 50,000 years. He had overcome both glacial and interglacial periods successfully and he occupied a territory spanning from Portugal to Central Asia. Despite his fabulous environmental adaptation, when Cro-Magnon (right) showed up, it took Neanderthal only a short time to become extinct. We can be certain there were crossbreeds as the Neanderthal people looked a lot like the Cro-Magnon. If fact we can believe Cro-Magnon mated with many of the other races around the areas of northern Africa all the way to Australia which was part of the Asian continent at that time.

Art

Magnon was crafty from the start and they have been found with numerous tools along with pieces of shell and animal teeth in what appear to have been pendants or necklaces. Great artists, works in ivory, and wall paintings were popular as shown below. They buried their dead intentionally showing knowledge of ritual and healed wounds show they protected family members.

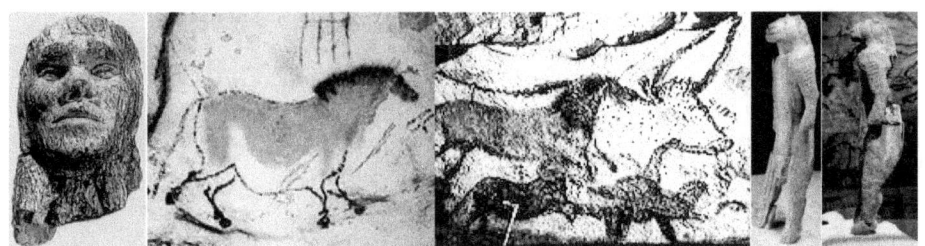

Cro-Magnon, Neanderthal, and the other homo-Erectus variants were not necessarily the best neighbors.

War and Extinction

Assyrian Description

The Assyrians told us about the Homo-Erectus variants and their interface with Cro-Magnon and the Annunaki [Anak] in "*Erra And Ishum*". This story provides us with a substantial amount of confirmation from other texts and is provided here in snippets. What we find in just about every ancient text is that during the Pleistocene, there were many wars. Some of them so horrible we cannot imagine the suffering.

6th Day Human-*Creator of the world, Holder of the sublime scepter, herdsman of the <u>black-headed folk</u>* [Homo-Erectus people], *and shepherd of mankind* [the Cro-Magnon people]. Here the Assyrians talk about the difference between the two humans. We know that Neanderthal was reddish-white and Cro-Magnon was white, we can believe most Homo-Erectus Variants were black.

Homo-Erectus Variants were hunted?-*When the clamor of human habitations becomes noisome to you, and you resolve to wreak destruction, "To massacre the black-headed folk."* The Homo-Erectus hybrids were to be massacred because of many things that upset the rulers.

Giants Wage War -*Lord Erra [the Creator], why have you plotted evil against the gods* [Anak people]? *"To <u>lay waste the lands and decimate the people</u>". Listen to what I say, as* concerns the people of the inhabited world, "I am king in the land, I am the fiercest among the gods. I am warrior among the Iggigi-gods [Angels], mighty one among the Annunaki [Anak people]." Identical to the Biblical Story, huge civil wars broke out between Hybrids, Anak and others. Each thought he was the mightiest. In this case, the angels were called Iggigi and the Anak were Annunaki, but the story is the same.

Assyrians described the Round Earth-*"Like the sun, I scan the circumference of the world-* Unlike later people the roundness of the Earth was well known.

Black headed folks again- "All the gods [Anak] are afraid of a fight, so the black-headed folk are contemptuous! These "Anak people" knew the trouble they could get in and stayed quiet, but the hybrid humans, evidently, kept fighting. While many texts blame most of the horrible wars on the Anak, this one says it was more the Cro-Magnon and Erectus variants that had learned the secrets of the Anak.

World Was Destroyed- *Indeed I left my dwelling and caused the deluge!"When I left my dwelling, the regulation of heaven and earth disintegrated:"The shaking of heaven meant: the <u>positions of the heavenly bodies changed</u>, nor did I restore them. The quaking of netherworld meant: the yield of the furrow diminished, being thereafter difficult to exploit. The regulations of heaven and earth disintegrating meant: underground water diminished, <u>high water receded</u>. When I looked again, it was struggle to get enough water.* Identical to other works, the Earth is flipped on its axis, flooded, shaken and disintegrated.

Greek Description

In Hesiod's book the **"Theogony"**, the Greek version of the history during the Pleistocene can be easily distinguished as shown below.

All the gods were divided *in strife, even to mingle storm and tempest and already hastening to make an utter end of the race of mortal men, declaring that he would destroy the lives of the demi-gods, that the children of the gods should not mate with wretched mortals, even should have living and habitation apart from men. But of those who were born of immortals and of mankind verily Zeus laid toil and sorrow upon sorrow.* [Like the Genesis story, the Angels and humans mated. The offspring were not liked by God.]

The seas acted horribly *and the earth crashed and heaven was shaken. Olympus reeled from its foundation as the undying gods charged* [War before the worldwide flood.]

A huge earthquake was felt on Tartaros *as the Titan's used their mighty missiles. The Titans launched the missiles against each other.* [Just like in the Jewish history, the War was enhanced by a Civil War. It was a pretty nasty war with missiles.]

The cry of both armies reached the heavens *as they met in battle.* [Satan and his followers try to take control of the planet Rahab/Venus, according to Biblical texts, but God destroyed the colonists.]

Zeus showed his mighty capability *as he hurled Lightning bolts. They flew thick fast and strong with thunder and lightning and a awesome, whirling flame.* [This could be referring to some type of Lightning Weapon.]

The Earth crashed and burned*. Vast forests crackled with fire all about. The hot vapor lapped around the Earthborn Titans. Unspeakable Flames rose to the upper air* ***[Outer Space]****. The flashing of the "thunder-stone" and lightning blinded their eyes of the Titans and the heat was tremendous.* [This was a nasty War for sure.]

The sounds were like that of Earth and the Heavens being smashed together. It was as if **heaven from on high were being hurled down.** [This sounds like a huge meteor storm that probably happened 11 thousand years ago.]

The winds brought earthquakes, *storms, thunder, lightning, and the "lurid thunderbolt"* [different than the thunder previously mentioned]. *The gods fought continually in cruel war. Three hundred rocks fell one upon another.* [This is a pretty good picture of the worldwide devastation from the huge meteorite show that occurred during that time.]

The gods launched missiles *and overshadowed the Titans. The Titans were* **buried beneath the earth**, *where they were bound with bitter chains. They were buried as far beneath the earth as Tartaros.* <u>*A fiery anvil fell down from heaven nine nights and days.*</u> *They would reach the earth on the tenth day. Again, A fiery anvil fell from earth nine nights and days and would reach Tartaros upon the tenth.* [This sounds like the huge meteor storm from Venus during its destruction 11 thousand years ago. I took 10 days to go between Venus and the Earth according to this history.]

<u>**Around the earth runs**</u> *a triple fence of metal like a necklace, while above the fence grows and unfruitful sea. There by the counsel of Zeus who drives the clouds The Titan gods are hidden under misty gloom, on Tartaros.* [The triple line might mean some type of missile defense system that thwarted attacks called the "unfruitful sea". The misty gloom of the Titans sounds like they were on Venus when it caught fire and became very hot.]

One thing I think we can safely say. The wars must have been horrible [The book of "Jasher" indicated that 1/3 of all the people died in the Pleistocene Wars.] We know Cro-Magnon had territorial conflicts with a wide variety of other races. The first race they possibly came into conflict with were the earlier

Homo-Erectus followed by Denisovan, and the Neanderthal, which had been around in the European continent for 50,000 years and had evolved from earlier populations, such as Homo Heidelbergensis. The sudden disappearance of the Neanderthal has been blamed on climate change and interglacial period adaptation but truly there were many wars and finally the Cro-Magnon triumphed. From ancient texts we can believe that the Anak got involved in a big way as well and the weapons used were not sticks and stones. To gain a better perspective of this "New Human" another entry into the Biblical book of "*Genesis*" might be in order.

Cro-Magnon of Genesis

First let me just say Haplotype scientists try to tell us Cro-Magnon humans were not related to Homo-Erectus or Neanderthalis before the Pleistocene Extinction 10 thousand years ago but it is obvious that only Cro-Magnon, Anak, and Mixed breed humans survived, and there were only 8 pure Cro-0Magnon, so most survivors were gentile mixed breeds with a wide assortment of homo-Erectus variant bloodlines and a little bit of Anak blood to finish them off. If you remember the 2008 study showed there were two types of humans we can believe there were two separate human lineages. One group starting with Homo-Erectus variants and a second line that mysteriously sprang up in the Middle East by some unknown method and mixed with Gentiles after the flood subsided. For years "undirected Evolution Theories" tried their best to meld some type of expansion disregarding DNA, but then we found out that if Homo-Erectus was the fabled 6th Age man and Cro-Magnon was the fabled 8th Age man/Adam, the story presented by Moses was unbelievably insightful and truthful. All undirected evolution ideas fall apart as they violate the _"Law of Entropy"_ which FORCES de-evolution.

The earliest known remains of Cro-Magnon are about 20 to 40 thousand years old, while Homo-Erectus claims ages from the tertiary period and there is little doubt that Homo-Neanderthalis came from Homo-Erectus, however there are strange mutations in Homo-Neanderthalis that made him more intelligent, stronger, and more Cro-Magnon like. We can assume these mutations were from the Anak people. All that being said, coming from the Middle East, Cro-Magnon and subsequent hybrids are found all over the world with a curiously small population in Southern Africa. Cro-Magnons were robustly built and powerful. The body was muscular; the forehead was fairly straight rather than sloping as found in Neanderthals and he had almost no brow-ridge. The face was short and wide and the chin was prominent. <u>The most interesting thing about Cro-Magnon is his brain capacity was about 1,600 cc; about 10% larger than either Homo-Neanderthalis or modern man.</u> [See Graph]

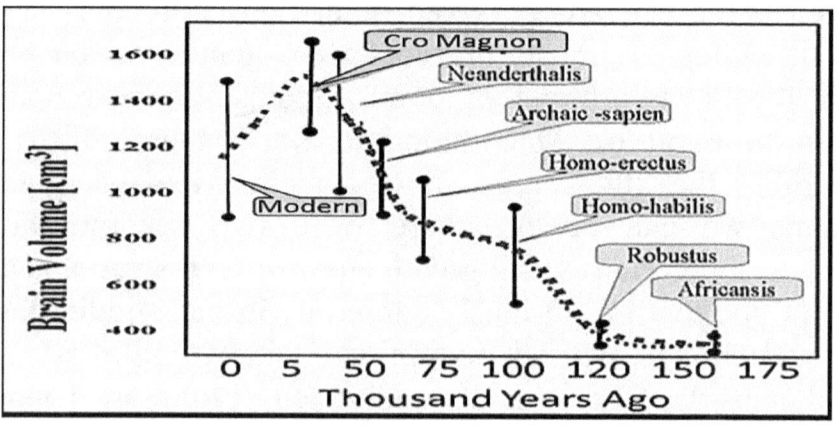

Later we will talk about what happened 5 thousand years ago to make our brain atrophy or shrink, but right now let's continue. Cro-Magnon was crafty from the start and they have been found with numerous tools along with pieces of shell and animal teeth in what appear to have been pendants or necklaces. Great artists, works in ivory, and wall paintings were popular. They also buried their dead intentionally showing knowledge of ritual and healed wounds show they protected family members.

For the next detail I must tell you a little more about Haplotyping. Essentially it uses mutation position and type to classify and time the various elements of an individual's heritage. As I previously developed, Cro-Magnon people are identified by Y-Chromosome DNA mutation [F] and mitochondrial DNA mutation [N] "Haplotype". This is written [F,N]. Homo-Erectus is identified as [A,L] and Homo-Neanderthalis as [D,C1] the last two type of people descended out-of-Africa. The simple Haplotype tree below shows lineages.

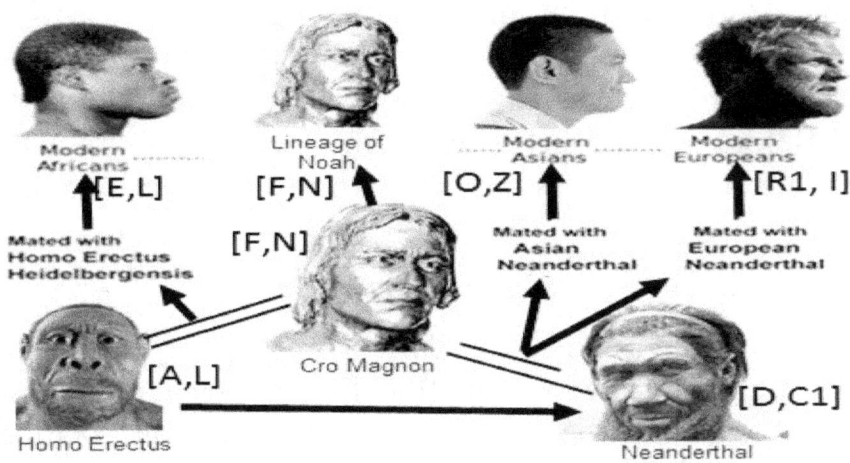

In a resent Haplotype study of almost 8 thousand individuals showed European lines had no A or B "homo-erectus" DNA mutations showing there was no early African link and almost no Neanderthal.

Bible Description

The Bible indicates a third human was created during the 8th Age. Scientists tell us Cro-Magnon humans simply appear one day at the beginning of the Pleistocene Age well after the Homo-Erectus humans of the Tertiary period. Neither science nor the Bible can separately answer us adequately, but together, the answer becomes clear. DNA tells us Cro-Magnon did not "Come out of Africa" so from where did he come? In fact, the researchers made note of their repeated absence stating "not one non-African participant out of more than 400 individuals in the Project tested positive to any of thirteen 'African' sub-clades of Haplotype A". The only remaining uncertainty relates to the identity of this "more ancient common ancestor". All that can be stated with confidence is that humanity's ancestor did not reside in Africa. Unfounded accusations of racism have become common as the prevailing Afrocentric hypothesis is constantly being challenged by the growing mountain of conflicting scientific evidence, especially in the evolving field of genetics. All Cro-Magnons are Modern, but all Moderns are not Cro-Magnon. Average stature of the Homo-Neanderthalis Man was about five feet four inches, whereas Cro-Magnon Man averaged over 6 foot. Before we can get a reasonable accounting from the Biblical history to help us out here, we need to sort of retime Genesis.

Retiming Adam

It is true that many Jewish and Christian groups try to explain that the Bible indicated Adam [the first Cro-Magnon] lived only 930 "years" and he must have been created on "October 23, 4004 BC at 9:00 AM". I have no idea if that was daylight savings time or not, but there are questions about that date. As I mentioned before, the Yawm/day word in Genesis simply means age, but the question is. "Can the time Adam was created be placed on the time that Cro-Magnon was created or were there 2 different creations? If we look at 2 different Jewish texts; Adam and Eve I and Adam and Eve II we find that Adam was possibly promised to live for another 5500 years after the convenient was made rather than the 930 years interpreted in the book of Genesis.

Adam and Eve I 3:2--Yea the word that will again save thee [Adam] when the five days and a half are fulfilled – the God in his mercy explained to him that <u>these were 5000 and 500 years and how one would come to save him</u> and his seed.

Adam and Eve I 21:9--*And [God] said to Adam, "O Adam all the misery which thou hast wrought upon thyself, will not avail against my rule, neither will it alter <u>the covenant of the 5500 years</u>.*

Adam and Eve I 38:2*- O Adam, as to the fruit of the tree of life, for which thou askest, I will not give it thee now, but when <u>5500 years are fulfilled</u>. Then will I give thee of the fruit of the tree of life, and thou shalt eat, and live forever, thou, and Eve, and thy righteous seed.*

Adam and Eve II 19:1*—Then God revealed to him [Adam] again the promise he had made. He explained <u>to him the 5500 years</u> and revealed to him the mystery of his coming to earth.*

Generational Dating

Knowing Jewish timing is generational rather than yearly and that early Years were actually called <u>"Jubilees" or 7 year time periods</u>, we simply use the relation rather than the specific date. If we simply convert years to Jubilee Years, we get the following chart. For confirmation Adam was told the 5500 year convenient time after Shem had been born so he was about 1000 years old by the chart following and the 6510 year death matches the details of the references provided. While this indicates Adam was born about 21 thousand years ago. It is pretty reasonable and it is triggered on the end of the Pleistocene age as the time of the worldwide flood and shift in the Earth which corresponds to other details.

Adamic Patriarch	Genesis son birth year	Son born year in Jubilees	New Birth [Yrs Ago]	Genesis death year	Died Age in Jubilees	New died [Yrs Ago]
Adam	130	910	20892	930	6510	15292
Seth	105	735	19982	912	6384	14333
Enos	90	630	19247	905	6335	13542
Cainan	70	490	18617	910	6370	12737
Mahalaleel	65	455	18127	895	6265	12317
Jerad	162	1134	17672	962	6734	12072
Enoch	65	455	16538	365	2555	14438
Methuselah	187	1309	16083	969	6783	10000
Lemech	182	1274	14774	777	5439	10000
Noah	500	3500	13500	910	6370	7130
Flood			10000			

This timing is significant in that many other pieces of physical evidence and written testimony indicate that Adam, or Cro-Magnon man, was created between 20 and 40 thousand years ago and this verifies that timing. If that was all we had I would not agree with it, but then we have the Biblical book of Ezra and many other similar dates by different societies.

4 Ezra 14: 10-12-For the world hath lost his youth, and the times begin to wax old. For the world is divided into twelve parts, and the ten parts of it are gone already, and half of a tenth part: And there remaineth that which is after the half of the tenth part. These writings come from about 2.5 thousand years ago. If we assume there is about a thousand years to go before the earth is destroyed and that makes up only 1/8 of the total time the earth has been around, the beginning of man would have been 28 thousand years before the time of Ezra or about 30 thousand years ago.

Babylonian History-*King Alalamar* [one of the first Babylonian rulers] *ruled for 36,000 years before the flood.* When added to the 9 thousand years since the flood, this king, ruled Babylon about 45 thousand years ago.

Manteo's "History of Egypt" - *Gods ruled for 13,777 years followed by 15,150 years of rule by demigods and spirits of the dead before the flood.* When added to the 9 thousand years since the flood, the first man or Gentile ruled Egypt <u>39 thousand years ago,</u>

Turin Papyrus- *Gods ruled Egypt for 13,420 years followed by 23,200 years of rule by demigods.* When added to the 5 thousand years since the papyrus was written, the first man or ruled about <u>41 thousand years ago.</u>

Emerald Tablets- According to this Egyptian text, *Thoth was sent to Egypt from Atlantis before it sank. He was sent 37 thousand years ago; he became the ruler; and at some point, he created the first 2 pyramids.* Assuming the 30 thousand years started near the time of Adam and Eve, and the reference is about 5 thousand years old, then Adam and Eve would have been here about <u>43 thousand years ago.</u>

Byzantine History- Syncellus wrote in the 9th century AD, that *the chroniclers of the pharaohs had recorded events for <u>36,525 years.</u>*

Note: It is true that these ancient references may be talking about Anak leaders, but the general reason for this section is to expand your view away from the strict 4004 BC timing.

Haplotype Verification

Besides the Bible, there are many texts that verify what we are told by Haplotype DNA studies. Homo-Erectus and Cro-Magnon were 2 separate creations.

Mandean/Essene- **"Iranian Folklore"**-*First God made man* [Homo-Erectus]-*Only later* [during the Bronze Age] *did he* [God] *put in a soul and teach him and make him erect. The soul entered Adam and he stood erect and talked and God taught him reading and writing and all knowledge.*

Yezidi/ Kurds-*"Mishaf Resh"*-God made man *[The Homo-Erectus had been made in the Silver Age]* <u>then</u> He *[God] created Adam and put in him a soul. On Saturday, He created Jibrail [a new Archangel to replace Gadrael/Satan] as a substitute for Tawus Malak. He made him head over them all.*

Gnostic-*"Hypostasis of the Archons"*- *-The rulers laid plans and said," Come let us [the Anak] make man that will be soil from the Earth."[Homo-Erectus man] They molded their creature wholly from the Earth. Now they made a body. They modeled that man after their own bodies. They did not understand the force of God. God breathed life into the man's face and he became a living soul. The man was called Adam. God made Adam sleep and opened his side like a woman.* In this version Eve was "attached" to Adam

Gnostic- *"Creation Text"*- *[The second Adam was Homo-Erectus].The third Adam is a creature of the Earth, that is the man of the law. He appeared on the eighth day/Age and became numerous and produced every kind of scientific information of the soul endowed Adam."* [We already read about the first 2 Adam's created by Incarnate God. This one would be Cro-Magnon. The 8th day is code for beginning of the Pleistocene.]

Haplotype DNA Mutation science indicated Cro-Magnon just appeared one day and other details indicate this was the famed Adam that Moses wrote about. The mystery of how Homo-Neanderthalis got so advanced from Homo-Erectus is a mystery to blind scientists, but it makes total sense once the Bible is used as a source of scientific information as there was a large number of scientists during the Pleistocene Age that the Judeo-Christian writers and others indicated worked on Genetics.

No One Describes 2 Human Types

If you are wondering why no one told you there were two types of modern humans, I can't say I blame you, but here is the really sad thing. There seems to be no studies about this obvious DNA characteristic oddity. I don't want to make a determination about these two types of people, but there are many in the labs testing DNA that know about this difference and they must be afraid to describe it for some reason.

Speaking of not telling you things in school, the next Haplogroup map shows the density of populations with certain mutations. I picked the X mtDNA Haplotype mutation for this example for one thing. The map shows that during extremely ancient times, North Americans came from a group in the Middle East somewhere near Israel and they never left. We can believe they were gentiles with Homo-Erectus, Anak and Cro-Magnon blood.

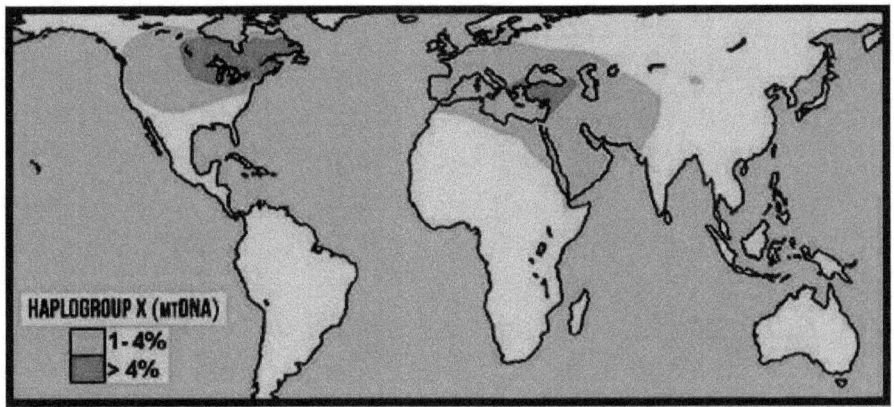

Now for the important part; they did not come across that stupid land-bridge from Asia. Besides this one example of teachers not telling students the truth or the information that could allow students to think, I can show others but they all say the same thing the X mutated Cro-Magnon came to America 10 to 20 thousand years ago in a direct flight from the Middle East. They did not walk and have babies along the way as they passed Asia into Alaska or anything like that. They must have flown here or they had great ships during the

Pleistocene. To help with all of this let's look at a little more DNA Haplotype tracking. The tracking seems to have issues in reasonableness, but we will make sense of it better in a later chapter.

Homo-Erectus During the Holocene

Right now we need to check out the track of mutation from Homo-Erectus. [$A_m:L_f$] DNA. What we will see is that this first human was in Africa 100 thousand years ago. Moved around a little for 10 or 15 thousand years and then a major mutation called Haplotype [$B_m:L_f$] occurred, but generally this person also stayed in Africa. At about the same time [$C_m:L_f$]. People with traces of Y-DNA Haplotype "A" or "B" mutations or the mtDNA "L" mutation seems to indicate a Homo-Erectus human was way back in their lineage, but do not misunderstand something important. No Homo-Erectus survived the Pleistocene Extinction. This means no Ergaster, no Heidelberg, no Denisovan and no Neanderthal. They all died during or before the Pleistocene Extinction. Only three groups survived, Pure Cro-Magnon, the Anak People and half-breed mixtures of homo-erectus variants with Anak, Homo Erectus variants with Cro Magnon or mixtures of all 3 subgroups [Homo-Erectus, Anak, and Cro-Magnon]. We know this because modern human DNA structures are all similar. If they had been based on Neanderthal, the alignments would be different. If you were wondering about the 2 different types of humans discovered in the chart 2 pages ago, I would have to guess the people carrying the Homo-Erectus bloodline most predominantly are one groups and those without this ancestor would show up as the other group-----Today, because of mixed breeding, there are no pure Cro Magnon and the Anak died about 3 thousand years ago. We are all gentile mixtures with Erectus, Cro-Magnon, and Anak blood. Even the Jews have married outside their lineage to become gentile. The entire book of Leviticus was made specifically to keep the original Jewish people isolated from everyone else [Gentiles and Anak]. It was impossible from the start. We are told that Ham son Canaan was the first to "stray" and Noah cursed him for it. These "Chosen Ones" had 612 rules that mostly addressed separation from the rest of the world. They did not live by them and soon all of the people of the world were "mixed gentiles". The problem with being a mixed breed is that our "spirit" is flawed and the only way to live in heaven after death is to get it fixed, but that is another story. As we all have Erectus blood of some type, I think a quick overview of Haplotype concentrations is in order.

Y-DNA Haplotype Tracking

As I said, by testing large groups, one can map out "generally" where individual groups came from, and how the lineage took control of various places around the globe. The following map shows a generalization of this type of Haplotype flow-map. Following the map are general descriptions of the various mutation grouping and a time-period for each event. The relative timing of the events is similar to known tracking, but I have compresses the timing so we can be closer to the ballpark. While we cannot get an exact time, we can determine what mutation comes first so we can adapt the sequence to known events. Hopefully, from my previous discussions it is known that this is not an exact science. There have been so many intermarriages and so many thousands of years, who came first, the Hamite or the Gaelic, can make it difficult to track entities and attempt the generalization of how a person got to be who he is today. Also, note that I confined most of the African mutations to Africa. While there certainly were extrusions, especially by the "E" grouping, these would not happen for a while.

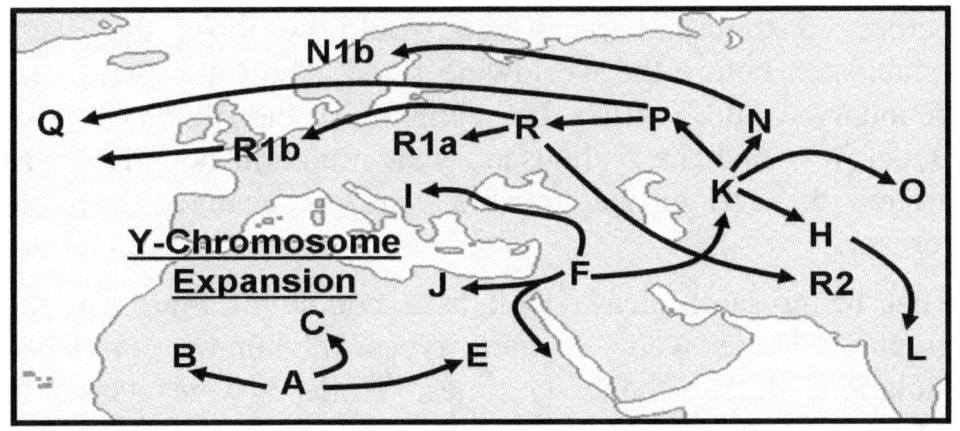

One can define major mutation or combination points by a letter and number identification. The first letter is the most significant ancestral point. It denotes MAJOR mutation points so we can time them pretty well. The number following indicates a single event mutation/modification and a second letter identifies an additional subset or combination of groups coming together. Please note the trail of "major mutation points" is characterized by general characteristics of the groups living in those locations and how they SEEM to flow. For the example above, the string "F to K to P to R" shows a particular timeline. This is believed to be the chain of ancestry for Europeans with F and K mutations occurring during the Pleistocene and "P" and "R" happening during the Bharata war 55 hundred years ago. Of particular importance is the Haplotyping known as R1b which is the base grouping of Northern Europeans and, strangely enough, Eastern North Americans. Sometimes names are given to these groups to make identification easier. The dates are approximations for reference and I have added the names.

It should be noted that some suggest that only the "A" Genome comes from Africa as the "B" type possibly started in a different country and descendants went "INTO" Africa. I don't know about all that and it happened before Cro-Magnon Man came along 40 thousand years ago. The mutation points before 10 thousand years ago would be those that were established before the worldwide flood. The 6 thousand year boundary would be from the beginning of the <u>Bharata War</u> or Babel War and the 5 thousand year mutation boundary was around the end of the horrible war that changed mankind forever.

That being said; please look at a couple of things in the graphic. There were 4 main types of humans or DNA mutations during the early times. While 3 were started in Africa [These would have been the Homo-Erectus base humans who intermarried with the Cro-Magnon], the 4th was located in or near the area known as Iraq [This would have been the Cro-Magnon people] . All of a sudden those living in Iraq mutated all over the place as if some massive nuclear event had overtaken them. As we go forward in time, we find that those staying near Persia mutated again while those who traveled to Greece and Africa remained less mutated. In Africa, the "C" mutation people wandered northward towards Egypt and were seemingly blasted to produce more mutations than other places. By this "Tree" it seems that civilization did not "come out of Africa". Certainly they were modified by the Anak people and sent other places, but walking to Australia probably didn't happen. As I mentioned before there is reference that the land of Nod [Hebrew for Wandering], where Cain and his descendents lived, appeared to be a melting pot of all the different humans who integrated with Cro-Magnon and possibly a place that sent out many ships to survive the ravages of the Pleistocene Extinction. Here are a few segments for the Jewish text "Generations of Adam" describing life during the Pleistocene.

Generations of Adam 19:1. *We proceeded to the Land of Cainan.-The inhabitants of the land are a mixture of the children of the Adam -and colonists from the Land of Haner. There are also some settlements of the descendants of Cain* *along the southeastern border of where Land of Cainan is located.— Colonists from the Land of Haner settled the Land of Nod after they had driven the descendants of Cain out of the land.-- In the seventieth year before <u>the Great Dispersion</u>, the people rose up in anger because of the plagues.--- You do those things that are forbidden because they bring only death and destruction, -and you would leave the whole earth desolate--- bringing in a flood on the earth. This flood will cover every manifestation of iniquity, washing the earth--- Twelve years before <u>the Great Dispersion</u>, Canaan entered the palace of King Coram and killed him on his bed--- the people of Abel were spared at the time of <u>the Great Dispersion</u>.-- people began to count their years before and after the Great Dispersion, for then all things were changed.--- Many were scattered in <u>the Great Dispersion</u>---- all traces of them [those practicing evil] will be removed from the face of the earth. Noah came out to warn the people of the Great Flood which will come on the earth, destroying all life except those whom God will spare.*

After the Great dispersion please notice that some landed in Persia. From there came the "J" mutation group to settle in Egypt. They were known as the Khemetians. Additionally, notice what happened as "P" quickly changed into "Q" mutation and was now found in the Americas as if by magic; unless one could have traveled by air or ship across the Atlantic to settle in what is now North America. I know that was hitting you with a lot of stuff so let's look at a different testing method as scientists check the Mitochondria.

Mitochondrial Haplotype Tracking

While you would think tracking mitochondrial DNA would show almost the same expansion and mutation. Generally, it is similar, but there are some differences. One reason might be that the Mitochondria are more protected so there are more "straight links". Along the left of the map shows the major MtDNA of the Americas. We will have to investigate how the N and M Haplotypings all of a sudden show up in America without secondary changes.

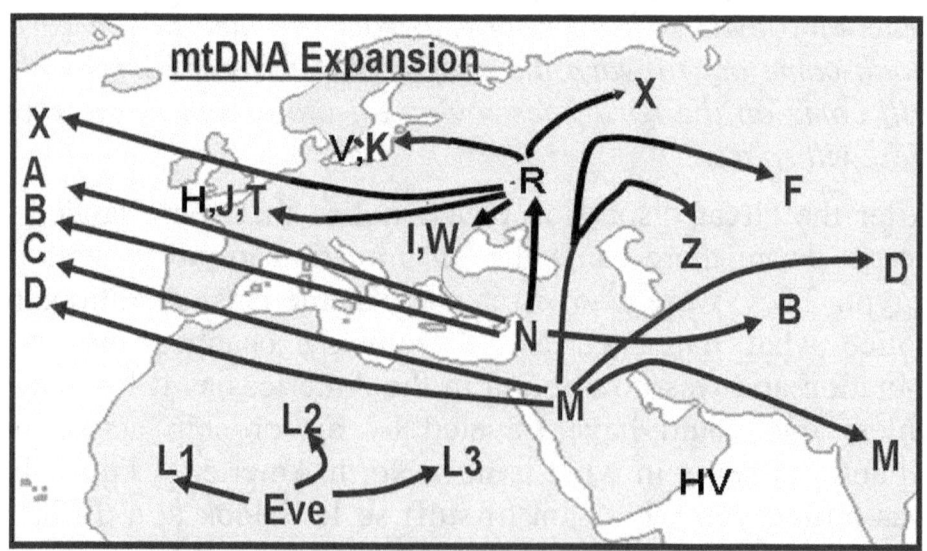

- The Homo-Erectus "L1", "L2", and "L3" lines stayed mostly Africa, but a group of those with this DNA became part of the Half breed Gentile groups that survived.
- The initial "N" Cro-Magnon line was all destroyed except for those who survived with Noah. Called the Chosen Ones, soon they married into the Gentile populations.

- The "M" Hamite and "R" Japhethite lines move as was described in Biblical texts Jepheth went north and married Gentiles that had survived. Hams group seemed to have had sex with just about anyone.
- The "R" mutation become the Scythian just as described in ancient Irish Histories and the "B" lineage is shown to push the "M" Dravidians to the bottom of India during the Bharata War as described in other histories and the "I", "W", "V", and "K" groups begin their westward takeover of Europe finally making the "H", "J", "T" lineages the main DNA mutations of the UK.

As with the Y-Chromosome Mutations notice that a large number of Mitochondrial mutations occurred 12 thousand years ago and another block happened 6 thousand years ago.

Homo-Erectus Humans
Eve=Erectus [100 thousand years ago]
L1= Sapien [50 thousand years ago]

Cro-Magnon Humans
N= Adamic [40 thousand years ago]
M=Arabic [30 thousand years ago]
L2= Negroid [20 thousand years ago]
L3= Nubian [12 thousand years ago]
R = Proto European [12 thousand years ago]
X= Proto-N. Amerindian [12 thousand years ago]
A=Adamic-Amerindian [12 thousand years ago]
D=Oriental-Amerindian [12 thousand years ago]
F=Mongol [6 thousand years ago]
Z=Oriental [6 thousand years ago]
B=India-Amerindian [6 thousand years ago]
C=Russo-Amerindian [6 thousand years ago]
V, K= Scandinavian [6 thousand years ago]
I, W = Greek [6 thousand years ago]
H, J, T=European [6 thousand years ago]

Travel to the Americas-From the map, I want you to see somehow, the X,A,B,C, and D mutation groups all got to the Americas without going through Asia. Many try to force the migration, but it doesn't make sense. They simply appeared. First the A and D groups followed by the B and C groups thousands of years later. It is a mystery, if no-one had flying transport machines. We will look at that later. Like the Y-Chromosome map, there are many variants of these flow maps as well. Again, it should be noted that these are my names rather than ones used by others. Please notice that there are not nearly, as many "Mutations" associated with Mitochondria DNA and the Y-Chromosome so we will concentrate of Y-DNA Haplotypings mostly.

Out-of-Africa Caution Again- There is a large group still trying to push the "E" mutation Haplotype [Y-DNA type] into all sorts of interactions, but there is little evidence of that. This "E" mutation is the "Out of Africa" idea. Variations may or may not have included mutation or significant mutation such that Haplographics could establish development and heritage of the Neanderthal, but understand something very important. The pure Cro-Magnon race had mutations 10 thousand and 5 thousand years ago and the Cro-Magnon/Erectus hybrids had similar mutations at similar times. I hope you understand by now that there were 3 distinct "beginnings" of mankind. One in Africa "L"Homo-Erectus-Cro Magnon], one in Persia "M" [Possibly Anak/Cro-Magnon from the land of Lod], and one in Syria "N" [pure Cro-Magnon] ---AGAIN. Nothing seemed to happen to the Africans with the mutation "L", but those living in Arabia and Persia mutated massively as if a massive nuclear bombardment occurred in those areas. Just before the end of the Pleistocene.

Noah and Important Events- Anthropologists completely ignore the worldwide flood that marked the end of the

Pleistocene. It makes no sense to them even though there has been a dozen studies showing the event happened, historical records from around the world [over 80 thousand stories and references are known], and many religious documents attesting to its certainty. I am not going to provide all 80 thousand flood stories; I just wanted you to understand that this thing DID happen. Key elements which greatly affected DNA and distribution of various races include the following.

- The creation of the Cro-Magnon or Adamic man appears to have happened 30 to 40 thousand years ago.

- The end of the Pleistocene Era was 10 thousand years ago.

- The worldwide flood associated with Noah evidently was 10 thousand years ago.

- Much physical and recorded evidence tell about the worst World War since the Pleistocene extinction. It happened 6 thousand years ago. We even know it ended 3100 BC.

- The real Aryan invasion described in many ancient texts from India, The British changed the dates to 1500 years ago to make themselves look good, but we now know it was about 55 hundreds years ago.

- The "Alien genes" found in Homo-Neanderthalis and PreInca skulls from 20 thousand years ago appear to come from Anak people.

- Unfossilized dinosaur remains showing at least some were living during the Pleistocene and the T-Rex and other bones show high levels of radioactivity.

- Nuclear Events [are recorded in histories, and physical evidence at two different times, 11 and 6 thousand years ago.

- Loss of enough nuclear material to run New York City for a year from the 16 strange nuclear processing plants found in Gabon, Africa. The beginning of processing Uranium now dates back to the Mesozoic times.

- Flying machines that could allow the mutation baselines found in the Americas to work are described in histories, religious documents, models, drawings, sculptings, and sky visible artworks since 50 thousand years ago or so.
- Ancient Jewish, Sumerian, India, Egyptian, and other ancient text, including our Bible, add information that should not simply be ignored.

All these things and a few other important things must not be ignored or the resulting theory of what Homo-Erectus humans really were won't make sense. Maybe we should look at another ancient text on Cross-bred Cro-Magnon and Homo-Erectus people we can call Gentiles. We find this detail in a Jewish Essene text.

"Mixture of Adam"

This fragmented text is now called "Mixture of Adam" and it c clearly shows there were all types of "races" during the Pleistocene. We can believe most of these changes were from inbreeding with the and Homo-Erectus variants, some may also have been mutations from the nasty wars going on around them. Interestingly there seems to be similarities between this data and the mutation characteristics found from Haplotype mutation testing.

11% Adamic Description-*4Q186 Fragment 1-* *His head and cheeks are fat; eyes are terrifying; teeth are different lengths; hands and fingers are thick; thighs are thick and very hairy; toes are thick and short.* <u>*His spirit has eight parts in the house of darkness and one in the house of light.*</u>

**[25%] Adamic Description-*4Q561 Fragment* - ** *His body hair is ample; voice is stern and does not strain; hair of his beard is plentiful; neither fat nor thin; short in stature; nails are* *strong...* [The percentage detail of this race and several of the others was not recoverable so it is listed as probable only by characteristic similarity.]

[40%] Adamic Description-4Q561 Fragment- *His beard is reddish; eyes are clear and circular; hair of his head ...*

67% Adamic Description-4Q561 Fragment- *His Head is wide; chin is thin; body is tall; body hair is full; build is thin but well built; hands and feet are medium length and thin; his eyes are fixed.* <u>*His spirit has three parts in the house of darkness and six in the house of light.*</u> *[67% Adamic]*

[75%] Adamic Description-4Q561 Fragment - *His hair is mixed and sparse; eyes are of a medium shade; nose is long and attractive; teeth are straight; beard is relatively thin; limbs in fit condition and medium built; elbows are strong and husky; thighs are of medium bulk; feet are of medium length; shoulders are medium width...*

89% Adamic Description-4Q186 Fragment -*His eyes are neither dark nor light; beard is light and curly; voice is soft and gentle; teeth are fine and well aligned; size is medium and well built; fingers are thin and long; thighs are hairless; soles of his feet and toes are even and well aligned.* <u>*His spirit has eight parts in the house of light and one in the house of darkness.*</u>

To make it easier to identify the differences, I have generated a chart of the characteristics from the fragments and identified how they relate to un-mutated Cro-Magnon or Adamic people. The percentages indicated in brackets were added to show presumed numbers because that was part of the missing information. They were determined by placement in the manuscript and details. The highlighted areas were either not recoverable or simply not mentioned. I have placed possible description generalizations in those positions.

% Adamic	11%	25%	40%	67%	75%	89%
Build	Thick	Med.	Med.	Thin strong	Med. fit	Med. strong
Size	Short	Short	Tall	Tall	Med.	Med.
Feet/hands	Thick	Strong Nails	Med.	Thin	Med.	Even/Thin
Hand length	Short	Short	Med.	Med.	Long	Long
Beard	Hairy	Hairy	Red	Full	Thin	Curly/Lt.
Body hair	Full	Ample	Red	Full	Thin	None
Head	Fat	Round	Round	Triangle	Oval	Oval
Teeth	Varied	Spaced	Spaced	Even	Even	Fine
Voice	Stern	Stern	Med.	Med.	Soft	Soft
Eyes	Scary	Round	Clear	Fixed	Med.	Med.
Nose	Fat	Fat	Med.	Med.	Long	Med.

While this type of characterization is skeptical at best, at least one can recognize that these early people recognized that there were mixed breeds of humans that possessed more of less of the Adamic blood. Noah's family were all full blooded Cro-Magnon [Haplotype "F"], but these other people were substantially different. If I were to guess Haplotypes of these people I would suggest the following:

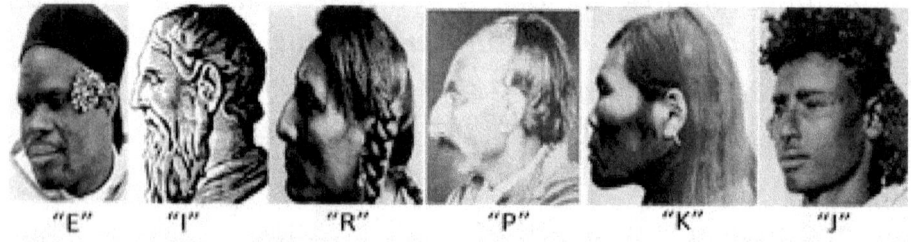

"E"　　"I"　　"R"　　"P"　　"K"　　"J"

11%-proto-Nubian [Haplotype E]　　67%-Proto Armenian [Haplotype G]
25%-proto-Greek [Haplotype I]　　75%-Proto-Asian [Haplotype K]
40%-Proto-Scythian [Haplotype R]　　89%-Proto-Canaanite [Haplotype J]

We really cannot guess how the mutations and variations occurred near the end of the Pleistocene. First people who lived during the Pleistocene were very civilized and professional similar to our geneticists, physicists, and engineers developed a high standard of living. The Bible and many ancient books go into great pains telling us about how most of the animals created in the Pleistocene were abominations to God as the DNA was manipulated. One group of the animals that were remade were some of the dinosaurs, another was a modification of human DNA producing some of the higher levels of apes. I could discuss a large number of texts and other evidence or you can just consider these as anomalies. The other things I mentioned were mutation and radioactive remains of some of the remanufactured dinosaurs. The ancient texts talk about massive wars and the radioactive remains suggests the use of some of the processed uranium of the prehistoric processing plants found in Gabon, Africa. While these both sound fanciful, the alternative is much more fanciful and everything is simply labeled anomaly.

We can believe that these early people recognized that there were mixed breeds of humans that possessed more of less of the Adamic/ Cro Magnon blood. With this information, let's clear up the Haplotype tracking map a little and look for the modifications near the end of the Pleistocene.

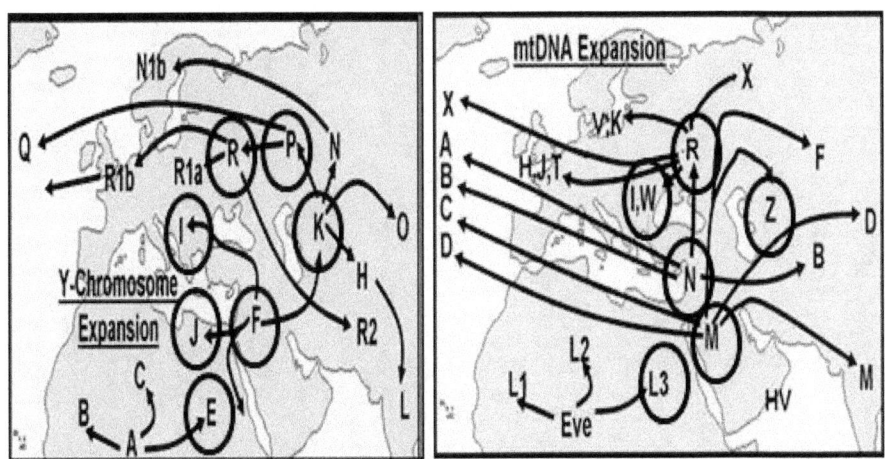

The Haplotype map of DNA mutation tracking shows where these first mutated humans settled for a time. The map on the left is the male Y-Chromosome mapping typically used as event separation is easier to discern, but the female Mitochondrial DNA map is shown to the right to show consistency. One of the problems with scientists accepting artifacts and DNA mutations that clearly show massive wars described in Judeo-Christian documents is that they are all over the world and those not reading Judeo-Christian literature have a hard time describing how the war could be worldwide. That's where Airplanes come in. Luckily, there are all types of physical and written evidence of flying transport and commerce. Possibly nuclear radiation aided in the modification of DNA.

Haplotype and Race

Certainly there is a strange correlation between DNA mutation and variations in race some of the traits are carried by the mt-DNA and others are supported by the nuclear or Y-Chromosome DNA. A general accounting of the basic race attributor DNA mutations is shown below. You will certainly find slight variations to this list depending on the group that makes up the description, but we are getting better and better at knowing when the mutations occur [except for the Mitochondrial mutations] and the basic location of each mutation [by using density of mutation in a group].

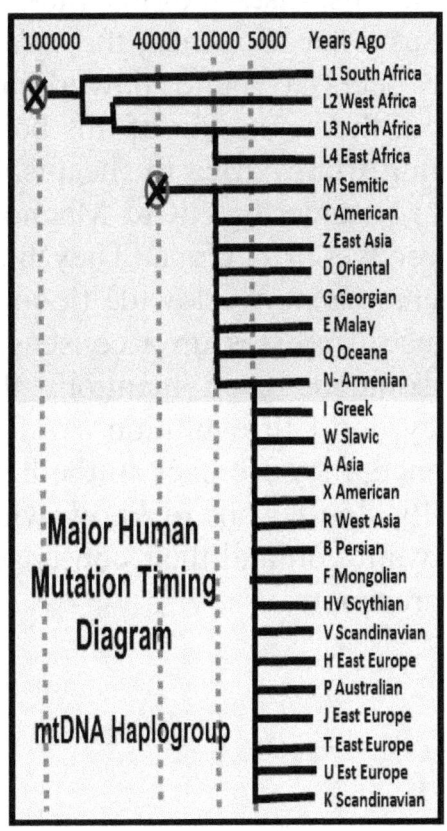
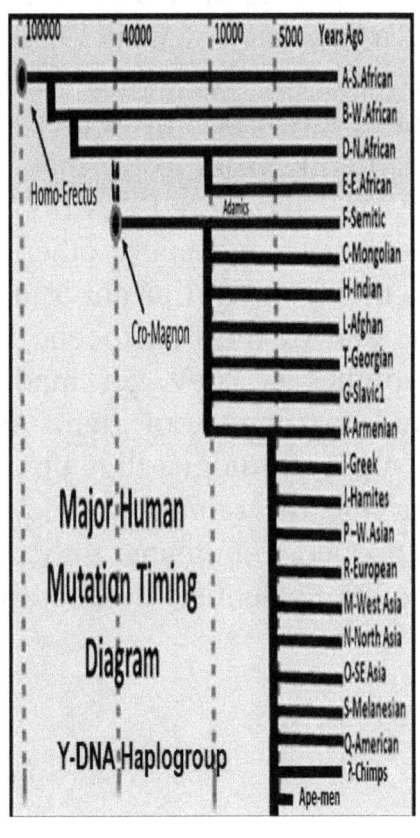

Like it or not, there is something strange about how the Homo-Erectus changed so very much when DNA mutation studies show that there were very few major mutations during most of the Tertiary and the entire Pleistocene. Also the mysterious Cro Magnon just pops into reality and is called the most perfect human ever by researchers. Others show that the Denisovan DNA has alien DNA and its highest match with modern man is in Australia. Many of these things I brought up at the beginning of the book, but the DNA only gets us so far and examining bones only gets us so far. If the Biblical descriptions make you uncomfortable, you can skip this next short chapter but I think it will help give you an additional level of perspective that seems to be missing from many anthropological scientists' suppositions. Instead of just picking up a few ancient works that tell us what happened, they want to rely on Consensus of their peers to build unworkable constructs. Like those presented at the beginning of this book vacillating from every man coming from Africa to sheepishly saying the predecessors of Neanderthal Cro-Magnon, Denisovan and many others never were in Africa. They huff up their bottom lips and state there was no worldwide flood at the end of the Pleistocene because it messes up a consensus hypothesis. They get mad if someone even mentions the predecessor race of giants as they don't fit evolution. As fast as a paleontologist digs up evidence, they are back in the hole burying the remains. They really don't want to know what happened to humans; they only want you and their consensus group of scientists to think they are smart.

Short Biblical Rant

Homo-Georgicus appeared to be more primitive than the Homo-Ergaster, but both had human qualities. Georgicus brain was about the same size as Homo-Habilis while Ergaster brain grew to 30% larger in the blink of an eye. One reason some might select Georgicus as the 6th Age human described in Genesis1:26 and many other ancient texts is that Homo-Ergaster was found in South Africa and Homo-Erectus was found in Southeast Asia placing Georgicus in the very center of the grouping, but I believe Homo Ergaster/Erectus was the first human made after the Cretaceous Extinction that destroyed just about everything.

***Genesis 1:26-31** - And Elohiym [This is plural for God] said, Let us make man in our image. So Elohiym created man in his own image-- replenish the earth—And the evening and the morning were the sixth Yawm/Age. [This was after a massive extinction when the Homo Gigantus had all died and became angels. Therefore Homo-Erectus was to REPLENISH the earth with humans outside of the Anak giants who were forced to return to the Earth after a great war in Heaven. Some suggest the use of US and our in this verse is describing the Anak making the 6th Age man.]*

A second reason some adopt Georgicus is that the Biblical testimony describes the creation of the 6th Age man in the first chapter of Genesis and in the very beginning of the next chapter it discusses the Garden of Eden believed to have been situated around the area of Iran which is just south of Georgia. That being said, the Garden of Eden would not be made until after the 7th Age so it could be interpreted as being thousands of years after the 6th Age man. Therefore, there was plenty of time for migration of Ergaster to the Middle East. If you don't mind, I think a quick Bible history insertion may help clear up some of the craziness of the development and sustainment of the Homo Erectus.

In Genesis, it talks about the human we are reviewing here being created during the 6th day after a horrible extinction that almost destroyed the whole earth with the Anak people being the only humans to survive. During the 7th Age, it says our Creator God Rested and watched his creation. Finally during the 8th age, he made a new human and placed him in a beautiful garden located at the beginning of 4 rivers.

Genesis 2:2-8- *And on the seventh day God ended his work which he had made; and he rested -then the Lord of the Elohiym/* [Creator God] *formed man of the dust of the ground, and breathed into his nostrils the breath of life; and* <u>man became a living soul.</u> *- And the* LORD *of the Elohiym* [Creator God] *planted a garden* <u>eastward</u> *in Eden; and there he put the man whom he had formed.* [Some suggest the reason The first chapter of Genesis talked about the Elohiym, plural of god, and Chapter 2 indicates the Lord of the Elohiym, created another man after the 7th Age was more special as it was addressed as having a "living soul". This would be the Cro-Magnon.

Where were the Eden and Nod?

One river flowing to this Garden of Eden [Delight] went west from Turkey [apparently along what is now the Black Sea] where it joined the others, Tigris and Euphrates and a third that included the Red Sea, which joined along the bottom of Turkey and would connect to the Black Sea River near the

Caspian Sea emptying in or near Iran. Very quickly let me just say, the 8th Age man did bad and was kicked out of the Garden requiring him to go farther eastward. Adam's son, Cain, killed his brother so he could marry his twin sister, Lebuda, and was forced to go even farther east to the land called Nod " The Wandering" somewhere in the Near East.

Genesis 4-16-*And Cain went out from the presence of the LORD, and dwelt in the land of Nod to the* **east of Eden**.

Cave of Treasures-*And she became pregnant and bore Cain and Lebuda. Then she became pregnant again and bore Abel and Qelima. Adam said let Cain take Qelima—then Cain said, I will take Lebuda for Lebuda was very beautiful"* [Lebuda was to be for Abel]

Adam and Eve II 1:6- *As for Cain, when the mourning of his brother had ended, he took his sister Lebuda and married her.*

Testament of Adam 3:5- *"A flood is coming and will wash the whole Earth because of the daughters of Cain, who killed your brother Abel out of passion for your sister* **Lebuda**.*"*

Jubilees 5:1- *And Adam begat thirty sons and thirty daughters and Adam lived 930 years.* That last verse has nothing to do with this story, but some may not have known about Adam's 60 children.

Cain Ruled Over the Homo-Erectus Variants

Anyway! Cain was told something very strange if you weren't reading about Homo-Erectus variants.

Genesis 4:6-7- *And the LORD said unto Cain, "--If thou doest well, shalt thou not be accepted? And if thou doest not well, sin lieth at the door. And unto thee shall be* **his** *desire, and thou shalt rule over* **him**.*"*

Some would ask *"Who did Cain rule over?"* He certainly did not rule over the Anak. They were more powerful, more intelligent and more organized. Assuming the 6th Age people were the Homo-Erectus and their associated groups, Rudalfensis, Ergaster, Rhodesinesis, Idualtu, Heidelberg, Denisovan, Peking, Neanderthal, Floresiensis, Grimaldi, Antecessor, and Boskop people found by the Scientists of today, they would be found around this Nod place. With Nod meaning the wandering, we can believe the Homo-Erectus people would be wandering in this area and find Cain living in his city named after his son Enoch. As you can see from the following map, showing the main places the remains of these people have been found, we can believe the Anak had taken these people all over the place and modified the original Ergaster as we have found. Cain could have certainly taken in many of these people into his kingdom. God's words seemed to say Cain would rule over the Homo-Erectus. As the Homo-Erectus variants all died out before the end of the Pleistocene, many struggle to understand how the world was repopulated after the flood and how unique animals were found in remote areas [Kangaroo, lemur, tiger, etc.] if the worldwide flood that marked the Pleistocene Extinction and Earth shift was so very bad. Genesis told us 5 times only the people and animals that were on the dry land perished.

- *Genesis 7:4* - *every living substance I destroy from off the face of the dry land/ earth.*
- *Genesis 7:21* - *All flesh died that moved upon the dry land/ earth- cattle and man*
- *Genesis 7:22*- *All in whose nostrils was the breath of life, of all that was in the dry land, died.*
- *Genesis 7:23*-*And every living substance was destroyed which was upon the face of the ground*
- *Genesis 7:23*- *They were destroyed from the dry land/ earth.*

Survivors of Extinction

We are also told the only pureblooded Cro Magnon or Adamic man that survived was Noah and his immediate family. Some of the 80 thousand very ancient "end of the world flood stories" tell us others found refuge by various means. Here is a tiny sampling. I am providing highlights for brevity.

Chaldean Flood-Survivors of the worldwide flood sent out birds to see when the land appeared. The gods [ANAK] also survived.

Sumerian Flood-The gods [ANAK] controlled the world before the flood. An ark was built and the seed of every animal was transported along with the human survivors. Birds were sent out by those saved from a worldwide flood. God also sent a rainbow afterward. The Annunaki [Anak]also survived.

Hindu Flood-Three worlds were flooded. Many were saved each time.

Bay of Bengal Flood-According to the India Indians, men grew disobedient. Puluga, the creator, sent a flood to destroy everything. It covered the whole land. Only two men and two women survived. That was the last time God and man spoke face to face.

Greek Flood-Greek gods [ANAK] ruled the land before the great flood which occurred during the 3rd Age of man. There were 9 survivors. The gods[ANAK] survived as well.

Celtic Flood-Giants[ANAK] controlled the world that ended by flood. 2 men survived the world destruction.

Lithuanian Flood-Because of sin and continuous war, a flood destroyed a world controlled by giants[ANAK]. Afterward, God sent the rainbow.

Egyptian Flood-Pyramid texts states –the 3rd Period was called the Golden Age of man and a worldwide flood destroyed it.

Egypt-In ancient Egypt, the Flood story was depicted as a boat full of animals as shown below.

Harappa-In ancient Pakistan, the Flood story was depicted as a boat full of animals. This is shown above right.

Southwest Tanzanian Flood-*The world began flooding, God told 2 men to go into a ship and take with them <u>all sorts of seed and animals</u>. The flood covered the mountains. After a while, the men sent out a <u>dove</u> to see if the water had dried, but it came back to the ship, then they sent out a hawk that didn't return and they knew the land was dry, so they disembarked.*

Lower Congo Flood-*The sun met the moon and threw mud, making it dimmer. While the moon was dim, a huge flood occurred. Men put their milk sticks behind them and were turned into monkeys. Later a new race of men was created.* We will see later that the milk stick thing is important with respect to devolution to Homo-Erectus.

Victorian Flood-*The Creator was angered because of the evil that people were doing, So he caused the ocean to flood by urinating into it. All the people were destroyed except those whom the Creator loved who finally became stars in the sky [Cro-Magnon had the ability to go to heaven] Besides this group a man and woman [hybrid Gentiles] also survived who climbed a tree on a mountaintop. From those two, the present human race descended.* In this story there are two types of humans that survived the flood just as we have been discussing. The urinating part is slightly different.

Chinese Flood-In the Chinese flood story, *the god named "Gong Gong" was ordered by the head of the gods to create a flood as a punishment for human misbehavior. The flood lasted 22 years, until the hero started to dam the waters. The hero was killed for this act, but from his corpse sprang a son who finished his father's project.*

Burmese Flood-*Nine people were saved from a worldwide flood and they sent out birds to test the land.*

Tibetan Flood -*Even Tibet was inundated with water until God, Gya, had compassion on the survivors and drew off the waters through Bengal. He sent teachers to civilize the people again. After the flood they were little better than monkeys. Again the reference to monkeys will be clearer as we discuss the Vanara people.*

Philippines Flood-*Two humans were saved from a worldwide flood.*

Maori [New Zealand] Flood-*Some people were saved from a worldwide flood.*

Totonac –Mexican Flood-*A man was warned of upcoming flood and built a boat. After the flood, he sent out a buzzard to test the land. The boat finally rested and God reversed man's face and hind parts and turned him into a monkey.* Remember the monkey part for later.

Aztec and Mayan Flood -*The 3rd Period was called the Golden Age. It was destroyed by storms. Nene [Aztec flood hero] was saved. He sent birds out to test the land.*

Yamana Flood-*The moon-woman Hanuxa caused the flood because she was full of hatred against the people, especially the men, who had taken over the women's secret kina ceremony and made it their own, so she let it snow so much that ice came to cover the entire earth. When it melted, it rapidly flooded all the earth except for 5 mountaintops. A few* *people survived on five mountaintops.* This is very similar to Biblical and geological evidence. People became evil; the ice age came; the earth apparently shifted or got warm all of a sudden; and the flood killed all but a few.

Andaman Flood-*The Supreme Being, Puluga, lived in the sky. Puluga created the whole world and man, but man began to forget his creator. Puluga became annoyed and sent a flood that covered the whole earth and wiped out the race. Four people escaped.*

Jicarilla (Apache) Flood *Before the Apaches emerged from the underworld, there were other people on the earth. Dios [God] told an old man and old woman that it would rain forty days and nights. People were warned to go to the tops of four mountains and not to look at either the flood or the sky [Gentile survivors]. The people didn't believe the old couple. When the rains came, only a few people made it to the mountaintops and shut their eyes. Those who looked at the flood turned into fish or frogs; if they looked at the sky, they turned into birds. After eighty days, Dios told the 24 people remaining to open their eyes and come down. These 24 people [Gentile survivors] went into mountains. Eight other people [Cro-Magnon] survived the flood. These were the ones that could travel because they could look to see where they wanted to go. These people told the Apaches about the flood before going into two mountains themselves. Around the turn of the millennium, the surface of the earth will again be destroyed, this time by fire.*

This story is special in that there were two types of humans that survived the flood. There were only 8 who could see where they were going just like the 8 Cro-Magnon that survived in the Biblical history. The other 24 Gentiles had to keep their eyes shut and trust that God would help them.

Blackfoot Indians Flood-*Some were saved from 2 separate devastating floods. The <u>thunderbird</u> was instrumental in their rescue.*

Hopi Indian Flood -*The 3rd Time was the time of destruction by water. The survivors sent out <u>birds</u> to test the land.*

Lakota Indians Flood-*There was worldwide destruction by flood. <u>Birds</u> were sent out by the survivors to test the land.*

Skagit Indians Washington Flood-*The creator made the world and 4 secret names. Only a few people should know the names, unfortunately, many learned them and became wicked. God made the world to flood. 5 people and some<u> of each animal type</u> survived.* This claim that the flood occurred because man had learned secrets he was not supposed to know is a common theme and should be considered as important.

Nizqualli [Washington] Flood-*The people became so numerous that they ate all the fish and game. They started to eat each other. They were so wicked that the changer god, Dokibatl, flooded the earth. All living things were destroyed except one woman and one dog, which survived atop Mt. Rainer. From them the next race of people was born. They lived like animals until the Changer sent a Spirit to teach them civilization.*

As DNA has told us, all the other survivors had Cro-Magnon attributes so they must have been those who had made a life with good old Cain and had crossbred. In "Generations of Adam" we find that almost all Adam's children left him and went with Cain so there were many Cro-Magnon to cohabitate with the other humans of the world. The map following shows the main areas these people have been found. Most died out even before the Pleistocene War. Neanderthal, Denisovan, Georgicus, Grimaldi, Peking man, and Floresiensis all were "developed" very near this land of Nod. Additionally, we find that "Homo Erectus variant" remains have been found all along the SE Asian lands including Australia.

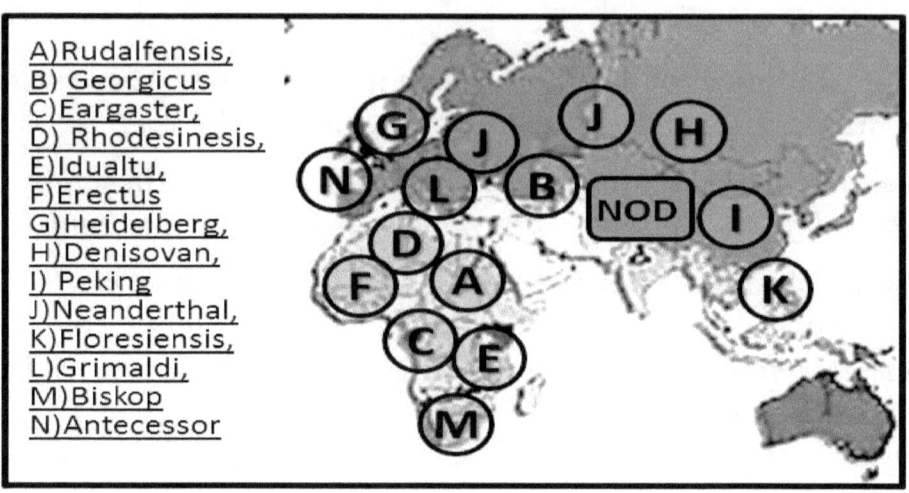

A) Rudalfensis,
B) Georgicus
C) Eargaster,
D) Rhodesinesis,
E) Idualtu,
F) Erectus
G) Heidelberg,
H) Denisovan,
I) Peking
J) Neanderthal,
K) Floresiensis,
L) Grimaldi,
M) Biskop
N) Antecessor

We are told Cain got a huge "family" and waged wars with his father's people. Apparently the Anak took some of the Homo-Erectus variants and changed them or they crossbred with the Cro-Magnon family of Cain and Lebuda. Prior to the Earth shift, the various populations of humans were located something like that shown next. These all died out except for the Cro Magnon and the mixing of Cro Magnon either by DNA splicing or human interaction.

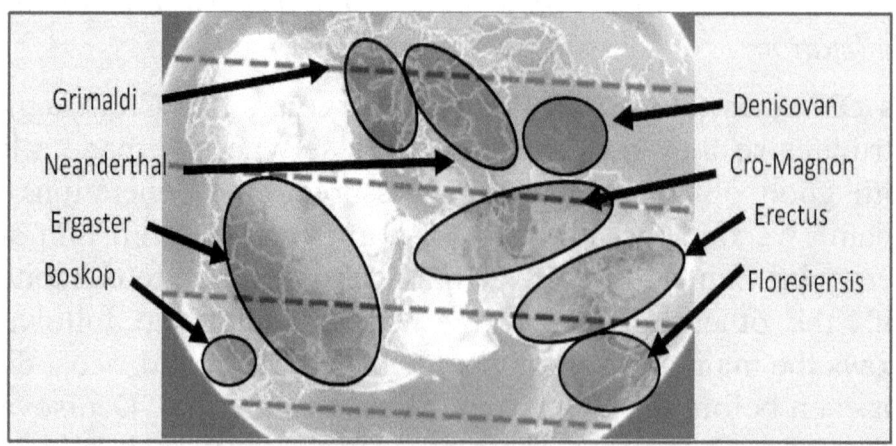

***Codex Junius II*-**Then the Anak began to take wives <u>from the tribe of Cain</u>, a cursed folk, and then the sons of men [Homo-

Erectus variants and Cro-Magnon] took wives from among that people, the fair and winsome daughters of the "Sinful Race" [Cain's offspring].

Wars broke out and some of the Cro-Magnon descendents had giant offspring after union with Anak people.

Generations of Adam 7:3- *The people of Timnor and Cain began to come upon our children* [Cro-Magnon or Adam's children] *and stole herds and produce, killing any who sought to prevent them.*

"Cave of Treasures"-*Adam knew his wife again, and she brought forth Seth, like unto Adam, and he became the father of the* **Giants** *[Anak]who lived before the Flood.*

Eliezar -*From them [the Anak] were born giants who walked about haughtily and indulged in themselves every manner of theft and corruption and <u>bloodshed.</u>*

Jubilees 5:1- - *the children of men began to multiply on the face of the earth and daughters were born unto them, that the <u>Anak saw them - and they took themselves wives of all whom they chose, and they bare unto them sons and they were giants.</u> And <u>lawlessness increased on the earth</u> and all flesh corrupted its way.*

God hated the animals that the people of that time "tried" to make and the continuous wars. He called the made up animals "Unclean" and decided to finally destroy most of them, including may different variants of humans.

Genesis 6:1-7-*The Children of men began to multiply on the face of the earth, - the sons of God [Anak] took them wives of all which they chose. ¨There were giants [Homo-Gigantus] in the earth in those days; and also after that, when the sons of God [Anak] came in unto the daughters of men, and they bare children to them, the same became the giants of old. ¯ And the* LORD *said, I will destroy man, beast, and fowl.*

Let me make a slight confession; there was a second group of Angels that became like the Anak. The Bible called them the Nephadim. Like the Anak they procreated with Cro-Magnon and later Homo-Erectus variants. During a Pleistocene War, all the Nephadim were killed and so were their offspring so I didn't think it was important---SORRY!

At the end of the Pleistocene, the Earth rotational Axis shifted, rain poured down and as the Polar Ice Caps melted and reformed, and massive unbelievable tidal waves covered the land. After the earth repositioned and the "new Earth" appeared, the following can be noted.

- Australian became an isolated Island that the liquid water level rose by over 400 hundred feet according to 12 different water height studies.
- The rivers of Eden became the Red Sea and the Black and Caspian Seas.
- Huge herds of Mammoths were quick frozen as Siberia now was in the Arctic
- A large number of the Anak also survived.
- We find that Cro-Magnon variants carrying Y-DNA Haplotype mutations of A, B, C, D, E, F, G, H, K, L, and T and mt-DNA Haplotype mutations of L, M, N, C, D, E, G, Q, and Z.
- While the first races of the Cro-Magnon based humans had been set before the Pleistocene Extinction, when the Earth shifted and the boats landed, these races spread out as described in the following Haplotype map.

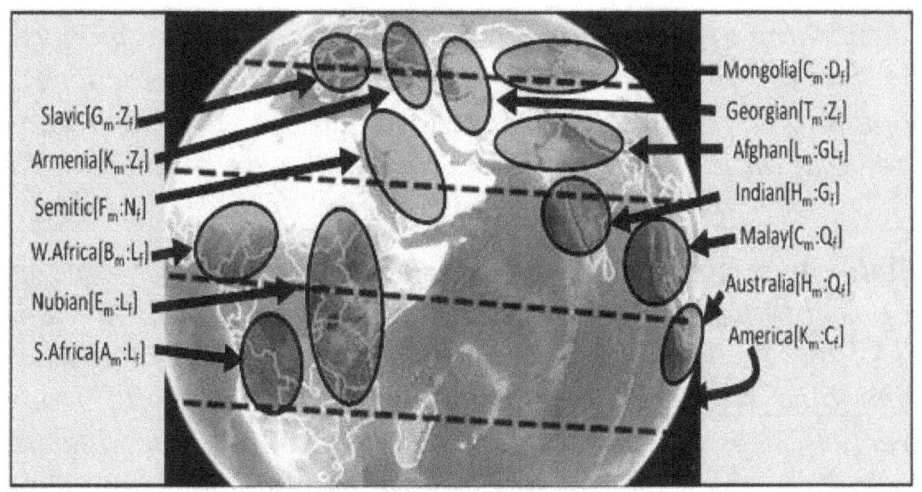

- The people with the Y-DNA mutations A_m, B_m, D_m and/or the mt-DNA mutations $L1_f$, $L2_f$, and $L3_f$ would have landed in Africa once the flooding halted. This group had not integrated with many of the others during the Pleistocene Age.

- The [F_m:N_f] survivors were the pure blood Cro Magnon Jews. We are told one of Noah's sons [Ham] had many half breed children that filled the land of Canaan with Anakim [descendents of the Anak].

- The [H_m:Q_f] survivors found themselves in Australia. One of this Haplotype ancestors had been of the group known as Denisovan

- The [H_m:G_f] survivors found themselves in India.

- The [L_m:G_f] survivors found themselves in the Near East.

A Large Study

Researchers from China, Indonesia, England and the U.S. collected samples, genotyped the Y chromosomes and analyzed the results. They looked for specific mutations at three locations on the Y chromosome and found that every one of the 12,127 samples typed, carried one of these three polymorphisms. –

> *According to their limited view, they indicated that modern humans of African origin completely replaced earlier populations in East Africa. When we add in the Earth shift, we see that all the Far Easterners died and were replaced with new families from the Land of Nod.*

Blindly they continued with their supposition. "*Only a couple of possibilities could go against this important information-If all of the Y chromosomes inherited from Homo-Erectus were <u>eliminated from the population</u> because those with Homo-Erectus ancestors were swept from the population due to a disease to which they were especially susceptible, they would not appear in the sample. Also, if only Homo-Erectus variant women mated with Cro-Magnon men, but no Cro-Magnon women mated with Homo-Erectus men, then there would be no Y-chromosomal evidence of the mixing.*"

Of interest, no Homo-Erectus remains have been found in China that are younger than 80,000 years old. Cro-Magnon arrived on the scene 40,000 years ago. While some indicate the climate caused the extinction of the Homo-Erectus there, before the Earth shift China would have been a great place to live. Possibly the Cro-Magnon didn't think they were pretty enough.

After the Earth settled 10 thousand years ago, there was a lull in human mutation then all of a sudden, massive mutations sprang up around the world. This happened around 5500 years ago.

More Mutation

As with many of the other supporting details, I'm not spending a lot of time on this subject but it, unfortunately, does tie in to Homo Erectus as you will see. The idea that ancient humans used and abused nuclear energy is distasteful to many so no one seems to want to talk about 2 of the worst wars we have ever had on our planet. The first happened about 11 thousand years ago causing about 50% of all major mutations on humans and the second one 5500 years ago caused the other 50%. Let's see what Geneticists tell us.

While for a long time there were not many mutations, the human genetic diversity today is vastly different from what 200 generations ago. A study dating the age of more than 1 million single-letter mutations in the human DNA code reveals that most of these mutations are of recent origin. Over 86 percent of the harmful single nucleotide mutations arose between 5 and 11 thousand years ago. Oddly, since then there have been few mutations at all. Overall, researchers now believe that about 81 percent of the single-nucleotide variants in the European sampled and 58 percent in the African DNA sampled arose in the past 5,000 years.

About half of the major mutation of the human race occurred 11 thousand years ago during the Pleistocene War and Extinction period and most of the rest of the major mutations occurred after the Bharata War.

Evidence

While there is tons of evidence of the use and destruction by nuclear materials during these volatile times, let me just point out a few.

- The 16 Ancient Nuclear Processing facilities located in Gabon, Africa is missing enough processed Uranium to power New York City for a year. ----our build bombs
- During the Young Dryas [11 thousand years ago, huge spike of radioactivity and other signs of nuclear event was records. The chart below shows some of the test,. The gray area would be the time just before the end of the Pleistocene [11 to 10 thousand years ago.]
- Many unfossilized Tyrannosaurs Rex bones not only show they lived 20 thousand years ago or so [a process of remanufacturing animals similar to that Jurassic Park movie] but also they are so radioactive, they must be painted with lead based paint to protect viewers.
- The City of Mohen jo Daro [mound of the dead] was left filled with the remains of bodies, melted walls and clay pots that had turned into balls of glass.
- Interestingly, in a sick way, the human remains scattered all over Mohen jo Daro are still radioactive.
- Around the world we are finding massive colonies of people who moved to underground cities 5 thousand years ago to protect themselves for something on the surface. Many underground cities were found in Turkey, Malta, China, Scotland, Mexico, Peru, the U.S.A, and just about everywhere else.

Many ancient text tell of the horrors of the war but I think the "Book of Jasher" gives a good indication of Mutations.

Jasher 7:19-20- *the name of the first son of Eber was Peleg, for in his days all the sons of men were divided, and in the latter days, the entire earth was divided. [in War] And the name of the second son was Yoktan, meaning that in his day the lives of the sons of men were diminished and lessened. [In this mutation a normal lifespan in excess of a thousand years was changed to less than 200 years.]*

Jasher 9:27-33 *And when they were building they built themselves a great city and a very high and strong tower; and on account of its height the mortar and bricks did not reach the builders in their ascent to it, until those who went up had completed a full year, and after that, they reached to the builders and gave them the mortar and the bricks [The huge citadel and; thus was it done daily. -let us[God's angels] descend and confuse their tongues---And from that day following,* <u>*they forgot each man his neighbor's tongue.*</u> *the Lord confounded the Language of the whole earth; behold it was at the east of the land of Shinar and its circumference is three days' walk..* [This mutation limited the capabilities of the brain. Therefore it began to atrophy, making our brains shrink from its 1600cm^3 to our current 1300 cm^3 over the past 5000 years.]

Jasher 9:35 *The 1/3 who said, We will ascend to heaven and serve our gods, became like apes and elephants. The 1/3 who said, we will smite the heaven, the Lord killed them, one man through the hand of his neighbor. The third division of men who said, we will ascend to heaven and fight, the Lord scattered them throughout the earth.* [One third of the Earth remained in a livable state, 1/3 died in the wars, and a third of all the earth mutated into what we will call the Vanara people]

Called the Bharata War by in the Indian histories, this time of massive mutation of all the people of the world was a horrible time not brought out in many classrooms. Some of the mutations were noted by major changed in how long people lived or how they could use their massive brain, but the worst mutation was more than simply establishing a new race of people called the Vanara.

Vanara People

Before we end, let me talk about something that is very curious and related to this whole evolution and modification of Homo Sapiens and Erectus. While not quite on subject, let's start out with what I had described before. Humans mostly mutated into races just before the end of the Pleistocene and a second time around 5500 years ago during what was called the Bharata War during which the book of "Jasher" tells us 1/3 of the population of the entire world lost their lives and another third became like Apes. The Indian ancient histories called them the Vanara. I know this sounds stupid so let's look at this and other similar texts.

Jasher 9:35-39 *-those who said, We will ascend to heaven and serve our gods, became like apes--*

Totonac- Mexican Tradition*-After the flood, the boat finally rested. Later God reversed man's face and hind parts and turned him into a monkey.* [This is probable an indication of man becoming ape-like during or after the war.]

Mayan Tradition*-During the second creation, people turned into monkeys and the world was destroyed by wind.* [This is probable an indication of man becoming ape-like during or after the war.]

Aztec History*-During the age of the four winds men turned into monkeys according to Codex "Laticano-Vatino"* [This is probable an indication of man became ape-like during or after the war.]

Lower Congo Tradition*-"First God created man. Sometime after a huge flood, men put their milk stick behind them and* *were turned into monkeys."* [This is probable an indication of man turning to a chimpanzee during or after the war.]

Tibetan History- *"Tibet was almost totally inundated by the flood. Later, the survivors had been little better than monkeys. The god Gya sent teachers to civilize the people and they repopulated the land after the flood."*

Around the world, all of a sudden, ape-men were provided a high level of respect in the communities. Images below show these ape-men protecting some type of electronic device and worshipping alongside the Egyptians. In India Hunaman was the revered "monkey-man" who was like a god to the Hindu. According to the sacred *"Ramayana"* He was a principle in the defeat of the dreaded Ravana. Hunaman did it with the aid of an entire ape-man army called the Vanara. Vanaras, including Hunaman, were born from humans, as if some mutation in DNA caused this anomaly. They spoke, wore normal clothing, worked with "normal" people, and battled the evil that continued to plague India after Vanaras organized into armies. After some time they died off as if their Procreation was almost impossible. In Cambodia many ape-men statues protect the Ankor-Watt temple. In ancient Mexico, one their sacred gods was depicted as an ape-man. On and on we could go, but soon the Ape-men were gone as the mutation must not have been sustainable in procreation. A Giant Hunaman statue is shown right.

By the way, just about everywhere you turn in Egypt you find depitions of one of these Vanara Ape-men. They were revered, and used in many setting as guards, and protectors of religious artifacts.

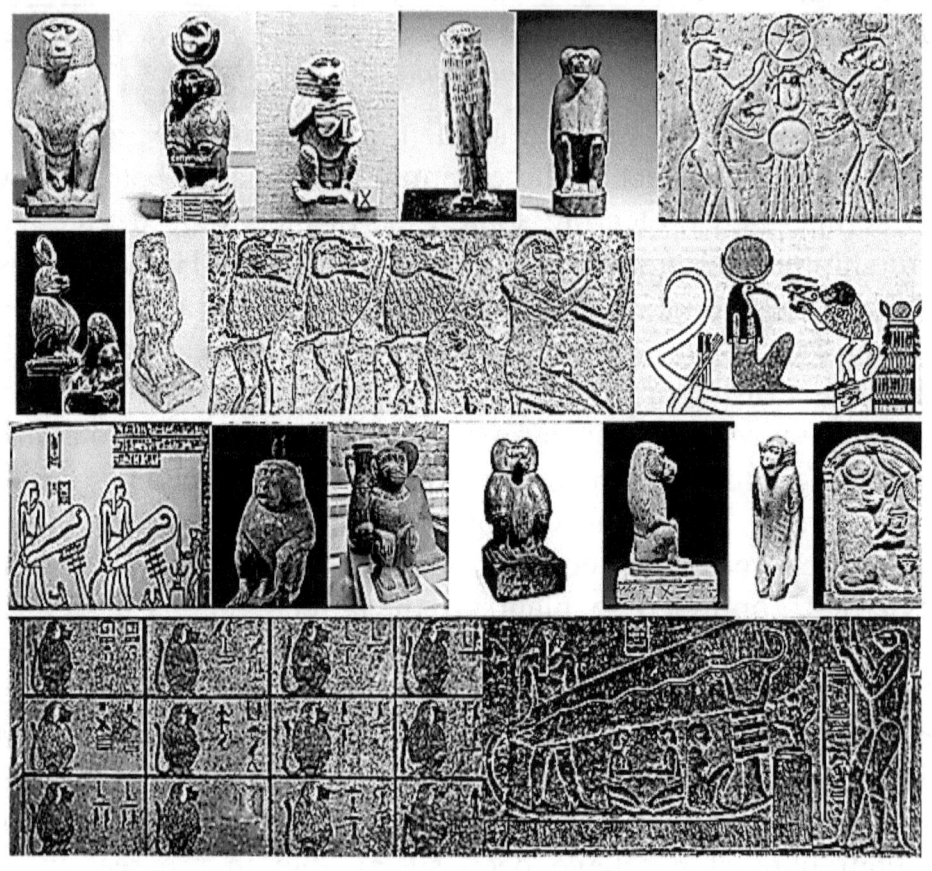

As the Vanara disappeared, Chimpanzee and Bonobo appeared.

Chimpanzees

I know this is not what you learned in school as you were told Chimpanzee and Bonobo evolved separately from man as both separated from the Gorilla lineage about 4 million years ago. Today we know differently. One thing is for sure, while the evidence keeps mounting that ape-men were useful members of society between 3000 and 2000 BC. Evolutionist scientists backed by consensus rather than fact will continue to lie, disregard, and weave complex reasons for ignoring evidence to hold onto their sacred religion over the ancient Judeo-Christian religious details presented. Besides many documented elements supporting the Vanara and the images, and the statues, and the many leaps of faith required to believe what you were told in school. While we can expect at least one of the Vanara mutations would have been successful in procreation, many scientist don't even want to think about it. Let's test some of the things we are finding out about Chimpanzee and its cousin the Bonobo to see if it makes sense chimps came from man and, most likely as part of the mass mutation that occurred 5500 years ago.

Test number One

A few years ago, one group of researchers studied the genomes of 12 species of Drosophila or fruit fly, four species of nematode worm, and 10 species of primate, including humans. By comparing with other groups of species, they were able to estimate how long ago the genes were likely to have been acquired. Rather than by evolution, they determined that a number of genes, including the ABO blood group gene, were confirmed as having been acquired by vertebrates through intrusion of viruses, protists, fungi, and Bacteria. They confirmed 145 genes were acquired by this means to shape humans. They found that 50 additional genes were "donated" in chimpanzee.

Test number Two

DNA structure of Chimpanzee is almost a complete match with humans. While one of the chromosome strings has been split in Chimpanzee, the makeup of the DNA sugar is very close. The following image is of a Human, mouse, and chimp X chromosome containing about 1,100 different genes, or sets of instructions. Each gene affects a particular trait in the body. Each specific nucleotide that makes up a DNA string [adenine (A), thymine (T), guanine (G) and cytosine (C)] show up as a slightly different shade in the image below. Notice that the Chimpanzee and Human are virtually identical. This is very strange. You can see how different even the mouse DNA is. Even the Centromere [necked down area] is the same in chimps and men.

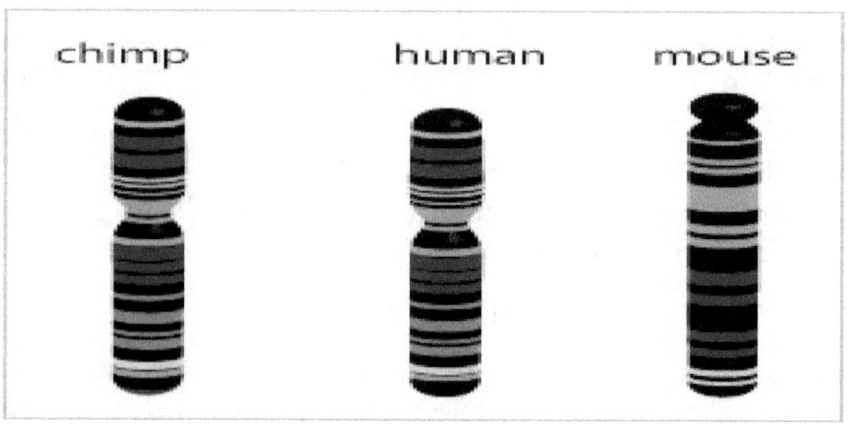

The following shows the SCB1,2,3 sequences of a Modern human, 2 Neanderthal, 2 chimpanzees [First five shown], compared to a gorilla, gibbon, baboon, rhesus, and colobus. A portion of the gorilla sequence had to be removed to make it match better. What we see is chimpanzee is so close it is obvious they are not descendents of the others.

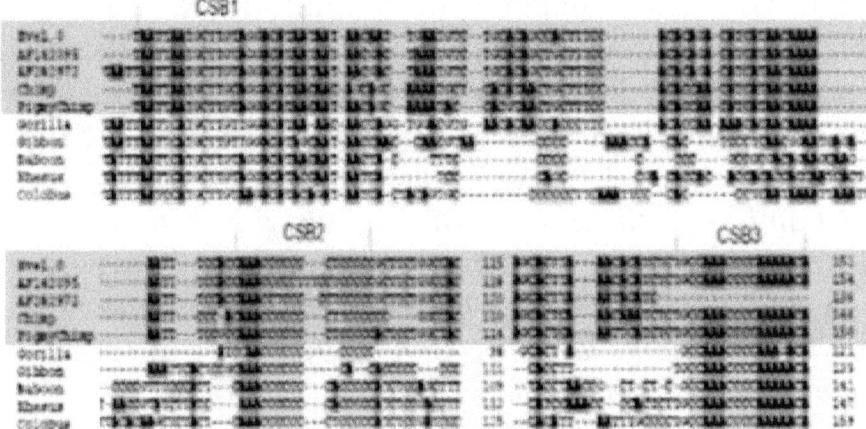

Test number Three

Human and chimp DNA was determined to be 1.2 percent different. Gorilla and other apes have over twice [3.1%] that much difference as they are somehow very different. In should be noted the genetic difference between individual humans today is minuscule – about 0.1%, on average. Now for the really weird part; bonobo has about 1.2% differences like chimps but Bonobo and Chimps have 1.6% difference between their DNA and Gorilla and other Apes.

Chimps and Bonobo are more closely related to humans than apes.

Test Number Four

At the end of each chromosome is a string of repeating DNA sequences called a telomere. Chimpanzees have about 23,000 base pairs of DNA that are repeated. While humans only have 10,000 base pairs of DNA repeats. One could determine that Chimps have not gone through as many mutations collecting these duplicates and is a newer species.

Test Number Five

It was determined that Bonobo was a mutated branch from Chimpanzee that occurred about a million years after the

Chimpanzee and Human split 4 million years ago [using nuclear decay timing]. This has been augmented recently. In three separate studies it was determined that the human chimp split could have been only <u>6500 years ago</u> making the Bonobo split only about <u>5 thousand years ago</u>. [In 2001*"Phylogenetic And Familial Estimates Of Mitochondrial Substitution Rates: Study Of Control Region Mutation In Deep-Rooting Pedigrees"*; 1997. *"A High Observed Substitution Rate In The Human Mitochondrial DNA Control Region"*. In 2000. *"The Mutation Rate In The Human MtDNA Control Region."*]

Test Number Six

The following chart comes from the journal *"Nature"* and has nothing to do with me or the seemingly crazy details I am trying to tell you about. The reason I put it here is I wanted you to look at where they placed the "appearance" of chimpanzee up in the right hand corner around the same time as modern man. While they wanted to show commonly discussed species they did not try to determine hypothesized lines of descent, just when we believe they "appeared". Disregard the times as I discussed previously as this is using nuclear decay timing still.

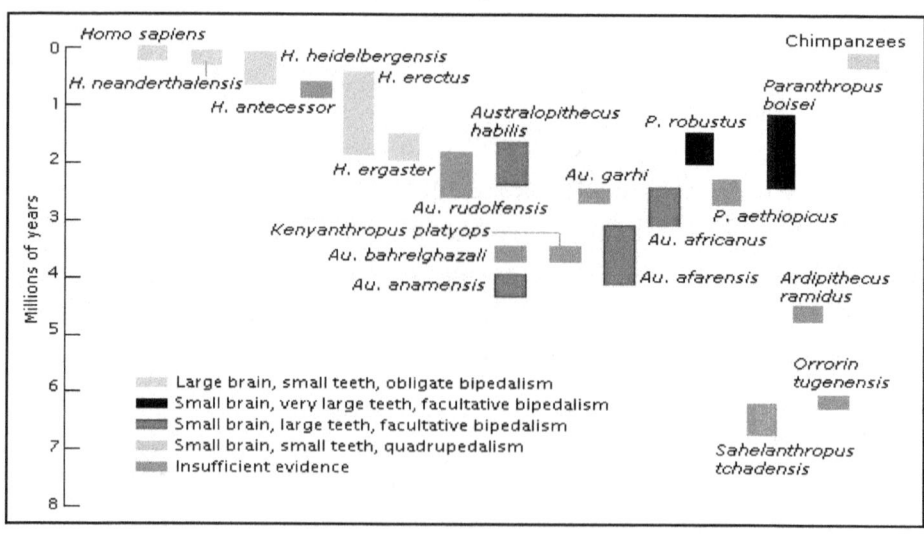

Test Number Seven

In the study, *"Bonobo Genome Compared With The Chimpanzee And Human Genomes"* Dr. Eichler and his colleagues found that the human and chimp sequences differ by only 1.2 percent in terms of single-nucleotide changes to the genetic code, but 2.7 percent of the genetic difference between humans and chimps are duplications, so we could really say it's only <u>1.1% difference</u>. They also found that more than <u>3% of the human genome is more closely related to either the bonobo or the chimpanzee genome than these are to each other</u>. They also found almost a thousand integrations of transposons [Transposed similar sequences] absent from the orangutan but present in bonobo, chimpanzee, and human. Of these, 27 are shared between the bonobo and human genomes but are absent from the chimpanzee genome, and 30 are shared between the chimpanzee and human genomes but are absent from the bonobo genome. The images below are of the Bonobo, Homo-Erectus, and Chimpanzee, showing what the small changes do to a person. In addition, about 25% of human genes contain "parts" that are more closely related to one of the two apes than the other. This suggests that Bonobo did not necessarily split from Chimpanzee. It could have been a new mutation of Human.

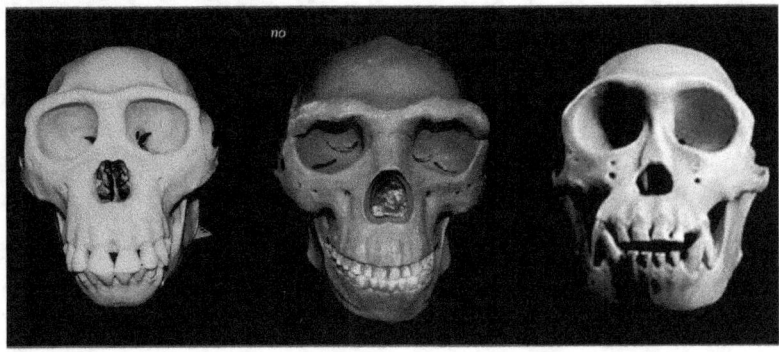

The following image compares Australopithicus, Erectus, Modern man, Bonobo and Chimpanzee skeletons. While the hips and hands have reverted back to Austropithicine style, there is a lot of similarity to the bonobo and chimp to homo-erectus. The bonobo is especially similar as noted from the skull and teeth similarity shown previously.

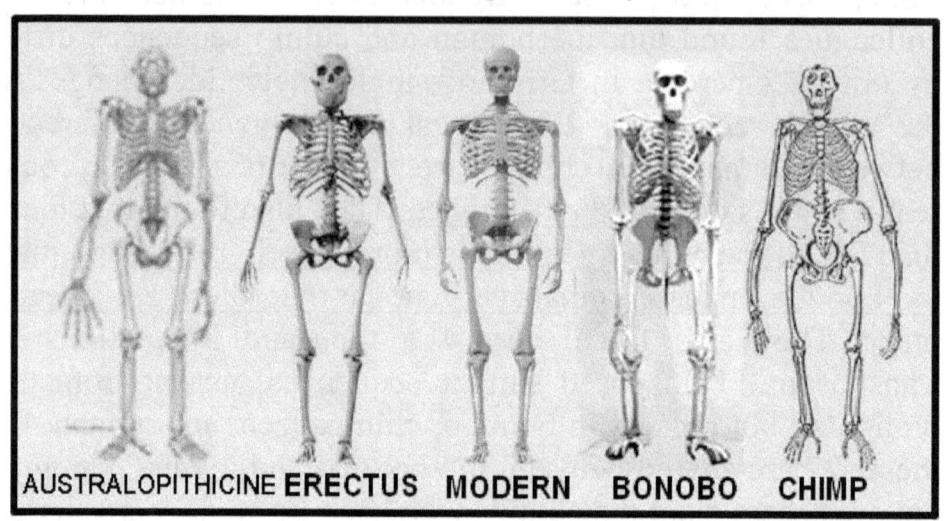
AUSTRALOPITHICINE ERECTUS MODERN BONOBO CHIMP

Sometimes Chimps are born with minimal hair as shown below. From these images we can better appreciate that chimps evolved from man. Yes their Arms are longer and legs are shorter and the pelvis is longer and wider, but chimpanzee seems to be a viable contribution to the idea of evolution as it looks like they evolved from us.

I'm not going to get into the horrors of this nuclear war in this book, but it seems we almost mutated ourselves back to Homo-Erectus by our stupidity 5 thousand years ago. The following description of the ape populations shows the strange bump in the road that produced chimpanzee. I corrected the timing according to the newer methodology as I would suspect some are not being taught this in our classrooms.

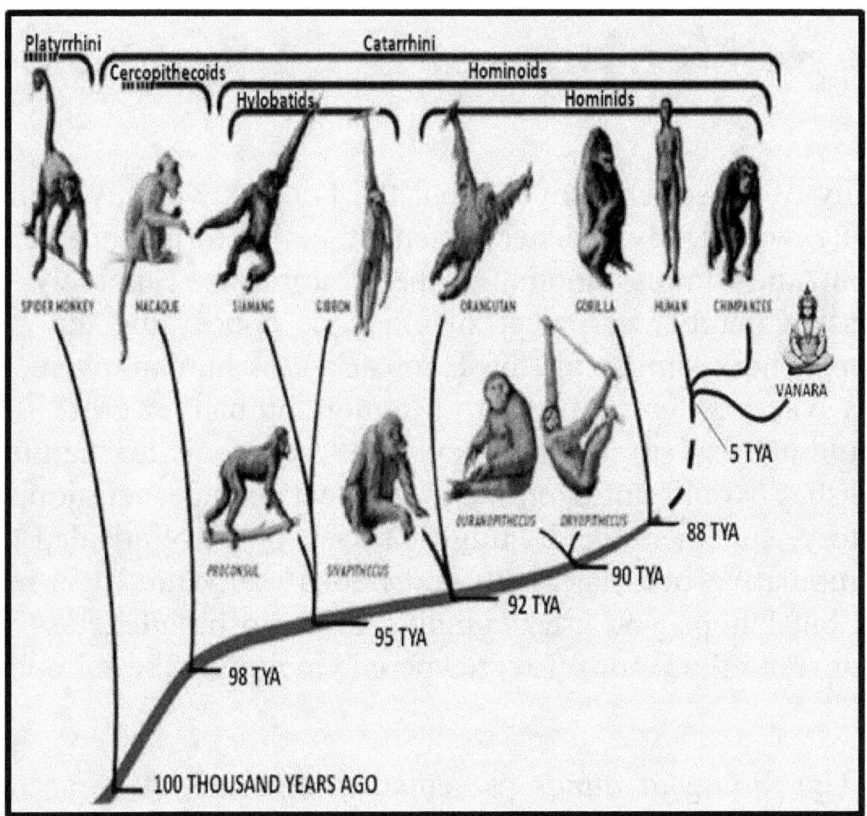

I'm sorry for the excusion away from Homo-Erectus, but I think this imformation is not being disemenated well in our classrooms and with that, let's recap and make some determinations.

Reflection and Conclusion

Many of those who haven't put the book down may wonder, "Why would any teacher, scientist, or historian not tell us about the timing anomaly, the chimpanzee anomaly, the massive number of pieces of evidence concerning the Anak people, how similar all the Homo-Erectus humans were, and how very different the Cro-Magnon humans were?" They would not describe consistency with the Judeo-Christian texts, and they would not even try to connect science, religion, and history together as it might show how blindsided and manipulative our classroom text books really are. It is really sad but I hope you are beginning to see a broader, less vain truth about the Homo- Erectus people including the following:

1. The timing of things presented to you cannot be accurate when they rely on the steady decay of nuclear bonds.
2. What we were told was 65 million years ago was more like 120 thousand years ago by 4 methods that have been cross compared for test and accuracy.
3. Giants walked with dinosaurs during the Mesozoic period. The giant people looked like us and were highly civilized.
4. Many texts including Judeo-Christina ones tell us the Giant people became the Angels of Heaven as they died.
5. Unfortunately for them, they wanted to continue having sex, getting their adrenaline racing, and all manner of things they could only do a humans. They tried to change heaven with horrible wars and they were kicked out and banished to Earth to become the Anak people.

6. Anything happening to Heaven happens to Earth as we are both linked together. With Heaven in shambles the Earth was identified as being a wasteland. This was the beginning of the Tertiary Period 120 thousand years ago.
7. For a timed, the Anak tried to make things work. They rebuilt towns and set up genetic laboratories where they began changing God's animals. They change the Australopithicus to the Sahelanthropus, to the Paranthropus, to the Naledi, and finally designed the Homo-Habilis, but he was still just a pet.
8. The Anak people cried for help so God made Homo-Erectus that was fashioned after the Homo-Habilis that the Anak had experimented on, but this guy had a much larger brain, could run, had normal hands and feet, was larger than Habilis, but he was hairy, and he could not talk.
9. Anak scientist went to work on Erectus to see what they could do. Some texts even indicated the Anak had sex with the lowly Homo-Erectus. All types of Erectus variants spread around the world to be servants of the Anak and to help them reclaim the land. Some were more successful than others.
10. Georgicus man variant was taken to the Middle East, but he was more of a throw-back.
11. Rhodesian man was a little more muscular, but still could not talk and he was left in Africa.
12. Java man and Peking man began making better tools and used fire, so they were getting somewhere.
13. Antecessor man was ok until they found he had a desire to eat humans. They left him in Spain.
14. Heidelberg man was big, but not too smart and they left him in Germany.
15. Grimaldi Man possibly could not talk, but he was sent to Italy as he seemed to be a great painter.
16. Idualtu man was big and brainy and was place in South Africa.

17. Floresiensis man turned out to be tiny so they placed him in an area with tiny elephants in SE Asia.
18. Denisovan man was sent to Russia. He was like a black Neanderthal
19. Neanderthal came along. This latest Homo-Erectus variant was pretty good. He could talk and had fairly good art skills and made moderately good tools, and was a generally good human, but the Anak still wanted more.
20. God made a brand new man called Cro-Magnon. This guy was wonderful. He could even sing and he was very creative.
21. Curiously, many people have none of the mutations associated with Homo-Erectus as they had no contact with humans who had lineages back to Homo-Erectus.
22. The Anak began teaching them everything including Genetics, Engineering, war, and sex.
23. God got mad and insured Cro-Magnon and Anak would not mate and have offspring.
24. Cro-Magnon could mate with the Various Homo-Erectus variants, and we can believe, the Anak tried their best to make partners available for the eager Cro-Magnon boys and girls.
25. Soon most people were half-breeds called Gentiles and they were very skilled at genetics and engineering helping to design aircraft and the reawakening of some of the dinosaurs.
26. War broke out and somehow Colonists on Venus got involved. The planet was destroyed and it may have be instrumentals in upsetting the critical balance of the Earth rotation and weather.
27. The end of the Pleistocene was expressed with a worldwide flood and a massive shift in the Earth's Axis.
28. Most animals and humans died, but some Cro-Magnon, Some Homo-Erectus half breed Gentiles, and some Anak people survived.

29. Just about all of us have homo-erectus DNA in us, unfortunately, we have none to compare.

Modification of Australopithicus to Habilis to Erectus to Neanderthal to Cro-Magnon and back to Bonobo and Chimp has not been by natural selection or uncontrolled evolution. It has been accomplished by hard work over the last 100 thousand years going against the law of entropy continuously trying to establish a better man rather than simply allowing Mother Nature to wreck havoc on our design.

The End

About the Author

Steve Preston is a long lime author of scientific, esoteric facts. His books focus on the painful truths rather than whitewashed details that make us comfortable. If you are interested in the truth instead of comfort, please review other works by Mr. Preston as shown below. The images are some from Egypt taking the older version of taxi. To the right the writer is shown in the Jewish Negev desert of Israel where the Dead Sea Scrolls were found. I found nothing but the marvels of Egypt and Israel.

Searching at Egyptian Pyramid   Searching in Israeli Negev

Development of Mankind Series
A Closer Look at Ancient History
A New View of Modern History
The First Creation of Man
The Second Creation of Man
The Antediluvian War Years
Bible Series
Adam to Abraham
Abraham to Moses
Bible Enhancements
Adam's First Wife
Closer Look At Genesis
Errors in Understanding
Expanded Genesis
Incarnations of God

Man After The Flood
The Creation of Adam and Eve
Six Deaths of Man
Twentieth Century to the End of Time

Moses to Jesus
Understanding the New Testament

New look at the Bible
The Devil
Why the King James Bible Failed
Old Testament Used By Jesus
The Antichrist

Tracing Cro-Magnon to Jesus
American Current Affairs
Allah' God of the Moon
Make Your Own Global Warming
Can We Save America?
Contemporary Issues
American School Disaster
Fast History of MILES Training
Monsters are Alive
Vampires among Us
America's Civil War Lie
Ancient History
Anak Gods
Ancient History of Flying
Behind the Tower of Babel
Driven Underground
Four Armageddons
History Confirmed By The Bible
Anthropic and Esoteric Science
Anthropic Reality
Awaken the Departed
Life Resonance
Releasing Your Consciousness
UFO, Astronomy, and Earth Science
Complex Earth
Creation and Death of Dinosaurs
Retiming the Earth
Victory of the Earth
Martians
Anthropology and Physical Science
DNA of Our Ancestors
God Didn't Make The Ape
Races of Men
Our 12-Dimensional Universe
Is Time Travel Possible?
Vibrational Matter
Irish-Jewish-Egyptian-Phoenician-Roman Connection
Mysteries of the Exodus
Mysterious Pyramids
Scythians Conquer Ireland
Truth About Phoenicia

Sex Crazed Angels
Promote the General Welfare
Terror of Global Warming

Great American Quiz
Our Very Odd Presidents
The Bad Side of Lincoln
Consensus Science
Humans on Display

Kingdoms Before the Flood
Lizard People
World War Before
World War with Heaven
World War Zero
When Giants Ruled the Earth

True Happiness
Self-Virtualization
Self, Soul, Spirit
Biophotonics and Healing

Not from Space
Where UFOs Go
Living on Venus
When Earth Exploded

Walk Through Time or a Wall
Meaning of Life and Light
Mystery of Photons and Light
Slip Through a Wall
Of Science and Religion

Moses Saved Egypt
Why Rome Fought the Berserkers
Truth About Hyksos Pharaohs

www.ingramcontent.com/pod-product-compliance
Lightning Source LLC
Chambersburg PA
CBHW071422180526
45170CB00001B/191